SPACE STORIES

ORAL HISTORIES FROM THE PIONEERS OF AMERICA'S SPACE PROGRAM

SPACE STORIES

ORAL HISTORIES FROM THE PIONEERS OF AMERICA'S SPACE PROGRAM

An Oral History Project conducted in conjunction with the Houston Chapter of the AIAA, Honeywell Corporation, and the NASA Alumni League

INTERVIEWERS

Dr. Robbie Davis-Floyd
Adjunct Professor, Department of Anthropology
Rice University, Houston, Texas

Dr. Kenneth J. Cox
Space Futurist

EDITING AND WRITING SUPPORT

Frank White
Space Philosopher

Space Stories: Oral Histories From the Pioneers of America's Space Program

First printed edition published in 2022 by Multiverse Publishing, LLC.

Print ISBN-13: 978-1-960119-71-1
eBook ISBN-13 978-1-960119-70-4

Cover images courtesy of NASA.

Book cover and interior design by Barbara Tada.
pixelgardendesign.com

Printed in the United States of America
2 4 6 8 10 9 7 5 3 1

MULTIVERSE
PUBLISHING
multiversepublishingllc.com

This book is dedicated to Guy Thibodaux,
father of solid rockets and an exemplar
of human potential.

CONTENTS

PROLOGUE

SPACE BOUND

If you're not willing to take a risk,
you'll never get off the ground!

—Guy Thibodaux and Max Faget

BEFORE HUMAN BEINGS had ever flown in space, they dreamed of it. But the necessary technical, political, economic, and spiritual forces had not come together in some 250,000 years of human history to create the conditions whereby humanity would break the bonds of gravity and fly out into the universe beyond. We were Earthbound. And then suddenly, we were so much more!

Official histories often make it appear as if nations make enormous decisions based on thorough research and understanding, after formal deliberations among "very important people" at the top of the heap. NASA's official early history has been told in countless institutional papers, reports, histories, and books. But a number of personal memoirs and histories have yet to become a part of the public record. When the individuals intimately involved in these momentous decisions are given voice, they tell a different story—one of hops and skips, of intellectual influence rather than personal authority, of personality clashes and chats among colleagues and friends, and of bootlegged designs that sometimes lead to billion-dollar programs. So it was with the birth of the National Aeronautics and Space Administration (NASA).

With funding from the American Institute of Aeronautics and Astronautics, Houston Chapter, from the Honeywell Corporation, and the NASA Alumni League, we embarked on the project of collecting oral histories from a number of individuals who were intimately involved in the events leading to the formation of NASA and the early development of the American space program. Our focus was not on the well-publicized astronauts, but on the inside stories of the engineers, administrators, and visionaries who worked behind the scenes.

Before Gagarin and Glenn—and before Armstrong and even NASA itself—men named Faget, Purser, and Thibodaux labored with passion and excitement at a place called the NACA—the National Aeronautical Committee on Aeronautics. Before astronauts climbed atop rockets and wedged themselves into minuscule capsules, the men at the NACA dreamed that they would, willing a space agency into existence. The workers at the NACA collaborated on more than mach speed and wind-tunnel experiments. They came together, for the most part, after World War II. As teenagers, they built and flew model airplanes—a passion and skill that eventually became a criterion for employment at the NACA. As government employees, they flew Frisbees off of balconies and through cramped offices crammed with gray desks to prove that blunt bodies have lift. They designed round rockets. They built gravity-defying couches for Yorkshire pigs to prove that humans could survive perilous reentries. They tried (and failed) to coax astronauts into spacecraft without windows. They were there at the beginning, creating NASA, spaceflight, and human footprints on the surface of the Moon. Thus our efforts with these interviews have largely been to publish first-person accounts by notable men and women who were intimately and powerfully involved in the events leading both to the formation of NASA and to the early development of its space program.

A word about our working process. With funding from the above-mentioned agencies, Robbie Davis-Floyd, anthropologist and space fan, and Kenneth J. Cox, former NASA senior engineer and space futurist, embarked on the project of collecting oral histories from a number of individuals who were intimately involved in the events leading to the formation of NASA and the early development of the American space program. The interviews we conducted took place in the homes of our interviewees and lasted three to four hours each. After transcription, Robbie and Ken edited these interviews for clarity and flow, and sent them to the interviewees for correction and additional comments. The final edited versions of these interviews fill at least 70 single-spaced pages each, and thus were too lengthy for book publication. (Interested readers can find these full texts by contacting Robbie Davis-Floyd at davis-floyd@outlook.com.)

The task of selecting the most relevant material in each interview to be presented in engaging and readable format was taken on by Frank White, author of *The Overview Effect: Space Exploration and Human Evolution.* In this task, he was assisted by Kris Stoever, daughter of astronaut Scott Carpenter, who donated her time because she believed so strongly in the importance of presenting these oral histories to the American public. In this volume, we present the oral histories of six fascinating figures, all key to the development of NASA and the American space program.

Guy "Tibby" Thibodaux, Max Faget, and Paul Purser were all principals in a NASA precursor agency called PARD—the Pilotless Aircraft Research Division. After the Russian launch of *Sputnik*, they quickly formed part of the nucleus of the space program. Caldwell Johnson, spacecraft designer, worked closely with Thibodaux, Faget, and Purser. We found a motherlode of historical perspectives in Robbie and Ken's interviews with Paul Dembling, General Counsel for NACA and principal drafter of the bill that became the National Aeronautics and Space Act of 1958, and Eilene Galloway, who worked tirelessly at the Congressional Research Services as a visionary policymaker and writer, helping to draft the legislation that created NASA in 1958 and the international space policy under which NASA has functioned since then. You, dear reader, can google all of these remarkable people and find a great deal of information about them on the internet, presented formally. These interviews were informally conducted and contain rich and funny insights and details that we are sure you will both learn from and delight in!

Our intention in presenting these six Space Stories is to show that institutional histories are, at root, histories of individuals who help to form teams working to do the near-impossible, and achieve their dreams. These interviews stand as a part of their legacy that is now accessible to the general public in easy-reading form—windows into the space-time continuum of the synergy and magic of their individual and collective contributions, inspirational stories from the past that also serve as cautionary tales for the future as humanity continues its long and exciting journey into the universe.

NOTE: Since completing the interviews and publishing the initial volume of *Space Stories* on Kindle, Ken Cox passed away. His work lives on in this book and in the hearts and memories of his family members and colleagues. He is missed and cannot be replaced.

ACKNOWLEDGMENTS

Our deepest thanks to Kris Stoever, daughter of astronaut Scott Carpenter, for her dedicated and time-consuming editorial work on the interviews presented herein, and to the American Institute of Aeronautics and Astronautics, Houston Chapter, the Honeywell Corporation, and the NASA Alumni League for their funding of the original interviews.

CHAPTER 1

GUY THIBODAUX
ROCKET SCIENTIST

BIOGRAPHIC INFORMATION

GUY THIBODAUX graduated from Louisiana State University with a BS in Chemical Engineering in January 1943. He immediately reported for active duty as a second lieutenant in the U.S. Army Corps of Engineers, and was assigned as a training officer at Ft. Leonard Wood, Mo. He was transferred to the China Burma India Theater and assigned to the 45th Engineering Regiment, built advanced fighter strips in the Burmese Jungle, and worked on the construction of a road from Ledo, India to Kunming, China.

After returning home, he worked at the Langley Memorial Aeronautical Laboratory's Pilotless Aircraft Research Division (PARD) in Virginia, where, in a period of three years, he became head of all propulsion and pyrotechnic activities at Langley and its launching range at Wallops Island, Virginia. He pioneered the redesign and modification of surplus military rockets to enhance the quality and types of aerodynamic data from free-flying supersonic rocket models and wind tunnels. He was responsible for the development of high temperature ceramic heated jets, electric arc heated tunnels, hypervelocity impact research, high vacuum technology, thermo-physics research, electron beam radiation, and oxidation resistant coating and thermal protection technology using ground and hypersonic reentry vehicles. He designed and operated an experimental solid rocket manufacturing plant and produced some of the highest mass fraction design spherical rockets from his own patented ideas, and developed novel manufacturing techniques. He conducted research on many solid fuel rocket operational problems that only occur in free flight. In 1958, he became a charter member of a committee that eventually planned the transition from the NACA to NASA and its first-years programs. He was instrumental in starting the Scout, America's only all solid propellant launch vehicle and the first one ever developed by NASA; he managed development of all propulsion and pyrotechnic systems on that vehicle.

In 1964, he became Chief of the Johnson Space Center's (JSC) Propulsion and Power Division and was responsible for all propulsion, pyrotechnic, and cryogenic storage and supply systems, power generation and storage systems and hydraulic systems on all of America's Manned Spacecrafts. He was responsible for the operation of large test facilities at JSC and White Sands, New Mexico, needed for the development, testing, and evaluation of these systems.

He retired in 1980 after completion of work on the space shuttle, and became a consultant to various industry and government organizations on such topics as solid and liquid rockets, free flight techniques, safety and hazards, and H-bomb simulation facilities.

INTERVIEW

This interview was conducted by Robbie Davis-Floyd and Ken Cox with Guy Thibodaux, the engineer and rocket propulsion scientist responsible for the propulsion work on Mercury and many other space projects, at his home near Johnson Space Center in Houston/Clear Lake, on Sept. 9, 1996.

Guy: The story hasn't been told about how the space program and NASA really got started. There are quite a few people still alive who are responsible for getting them going. For example, I've never seen any references to the fact that some of us spent a lot of time up in Washington, going through the Pentagon like we owned it, coming up with what became the space program. It's not documented, you see. What I am really interested in is covering that gap—the transition between the NACA and the events leading up to that, up to the early years of the Space Task Group.

Robbie: Before we get into the space program, tell us a little bit about yourself. Where were you born?

Guy: I was born in the Louisiana swamps, at the F.B. Williams Lumber Camp in the Atchafalaya swamp on the west side of Lake Verret. It was a big cypress logging organization. My father worked there. My birthplace was registered as Napoleonville, Louisiana, which is 12 miles north of Thibodaux, Louisiana on Louisiana Highway 1, which parallels Bayou Lafourche.

Robbie: Is that town named after your family?

Guy: Yes.

Robbie: Did you grow up there and go to school there?

Guy: No, we left there and moved to New Orleans when I was about 5. I went to high school in New Orleans, and later on I went to Louisiana State University. The interesting part of it is that Paul Purser, Max Faget, and I were all LSU graduates. Max and I were college roommates. Max and I had a pact that at the end of the war, if we both survived, we'd get together and go look for a job together.

Robbie: What was your role in the war?

GUY: I was an officer in the Corps of Engineers and I served in the China-Burma-India Theater building a road from India into China and some advanced fighter strips.

ROBBIE: What was Max's degree in?

GUY: His degree was in Mechanical Engineering with an Aeronautical Engineering minor. Paul was the same way as Max—they were both in aero-engineering, mechanical engineering. Paul graduated about two or three years before we did.

I know Max as well as I know anyone, and I knew him longer than anyone. When I met Max, he was a transfer from San Francisco Junior College. He was already a sophomore, but he was a freshman in Military Science. I was his platoon sergeant, and I thought he had two left feet! That's how I first met him. Later on we got to know each other fairly well. I'm not sure why—we were both about the same size, and we're both from Louisiana, but he's not a Cajun, I am. I had to work to earn some of my school expenses, and it took me four and a half years plus a summer school session to finish, because there was one required course I had to have that I could not schedule because of my work. During the summer school session, Max and I roomed together. Then I had another semester that I had to finish so we roomed together again.

There is an interesting story about how Caldwell Johnson—a technician who worked closely with Faget on spacecraft design for many years—got hired. The way they hired people at Langley was they went to these model airplane meets because they knew that the kids who built the best model airplanes were the best craftsmen. When Langley was hiring craftsmen, they would check up on these kids who were winning all these model airplane meets. They would find out the kids who were building the best models, and those were the ones they would try to convince that they ought to apply for a job to work in the shops out at Langley.

So, they convince Caldwell he ought to apply for this job. To meet the requirements for the job, he couldn't be color-blind. So Caldwell was rejected by the doctor and he went in with his rejection slip, whatever it was the doctor gave him, and talked to Mr. Sharp with his tail hanging down between his legs and his chin down and said, "Mr. Sharp, I don't qualify for a job. You can't hire me." Sharp said, 'What the hell's the matter, son? He said, "I'm color blind." Sharp looks at the map on the wall and points over to the Pacific Ocean and says, "Son, what color is that?" Caldwell told me that everybody knows the ocean is blue, so he said, "Blue." He said Sharp pointed to another ocean and said, "What color is that, son?" "Blue." He said, "Hell, you ain't color blind, son. That doctor doesn't know what the hell he's talking about. You passed!" And he scratches out the rejection and said, "You're hired!"

You could never do something like that today. Caldwell didn't even finish college, in fact. Here he is, this designer of American spacecraft, but he never finished college. He went a couple of years, maybe. He went through the apprentice school. He was one of the top students in the class, so he became an engineering draftsman. Then he headed the section that designed all the little models we used to fly out on Wallops Island.

KEN: Apparently model building, in those days, was really an area where a lot of people became craftsmen.

GUY: Yes, and there is a story about Max that is priceless. When we first went up to Langley, Max was still a big model airplane builder. His friend Woody Blanchard eventually became the one of the top model airplane builders in the United States for a number of years. We'd all go to meets together. I was what they called the "gopher" ("go-for").

We were going up to the New York Daily Mirror Flying Meet at the Grumman Airfield in Bethpage, Long Island. The back of the car was filled with model airplanes. Max and Woody and I went to eat breakfast on the main street in Newport News, Virginia. At that time in the morning the street was totally empty. This car is a Chevrolet Coupe and it's canary yellow, just as bright as can be. We walked over and had breakfast. Woody and I reached into our pockets and said, "I got the exact change, Max. You don't have any change, do you?" He said no. And we said, "Well, we'll leave it and you pay for the bill. We are going to go back to the car. Meet us there." The car was parked right across the street.

Woody and I went and sat in the car. No Max, No Max. No Max. No Max. *(laughter)* Pretty soon we look around, and here comes Max walking right past the car. Woody starts to holler, but I said "Shhhh!" Max walked right past this canary yellow car. He walked all the way to the other end of town and came back. Then, when he came back and it had gotten a little bit lighter, he looked at us with this big grin and said, "Where you guys been?"

He just had his head in the clouds. He was thinking about the meet or something. He was that way about a lot of things. When we went to Langley, it was almost 6 months before he could find his way to work. I used to have to tell him how to get to work every morning. *(laughter)*

KEN: What was it that attracted you early to a friendship with Max?

GUY: We didn't have too many common interests. We did share interests in sports. We played a lot of handball together. We were both very similar students. We didn't take our college work that seriously. *(laughs)* We were pretty independent. No one told us what we had to do. The professors couldn't manage us very well. If we got a lousy grade, it was usually because the professor didn't like us and he couldn't manage us, or that we just weren't interested in the subject. My whole attitude was I was there to learn something, not there to get a fancy grade. I knew a lot of people who got fancy grades but who didn't know a damn thing except they memorized stuff by rote and they could answer the test questions. I think that with our school records, they'd have never hired Max and me today. We couldn't even qualify to be hired over here [at JSC] today. Neither one of us were honor society types or anything else. We flunked out more roommates than anyone I know because we never studied. I never took notes in college. I'd come home at night and I'd try to remember what I had heard and I'd do all my note taking at night. If I couldn't remember it, it obviously wasn't very important.

Ken: You must have a pretty good memory.

Guy: I had a pretty fair memory. Somewhere along the way, I understood what I was good at and what I wasn't good at. I was fortunate enough that the profession I took up was one which I was very happy I chose. The teaching methods they used in my department were better, far better than anything else I've seen. We were taught to *think*. The problems we had to solve were thinking problems, not multiplication and addition and stuff like that.

We were also taught teamwork. All of our labs were done by teams of people. We rotated the leadership and somebody had to figure out how to adjust all the valves so the thing would flow in the right direction. Somebody had to collect the samples. Somebody had to weigh them. We had to give assignments.

The old department head made us think. He was a really great teacher, very demanding. The other guys who told me I had to keep this notebook and I had to be neat and look this way—

I didn't pay much attention. *(laughs)* Max and I are really not that great for neatness.

Robbie: *(laughing)* What did your dorm room look like?

Guy: When I was a cadet it looked great, because I would have been walking drill tours if it wasn't. It was spit and polished and shined and the bed you could bounce quarters off of and the whole bit. But after I got out I got out of that—well, it was sanitary. It was a hall, that's what it was. We lived in a stadium. They had cheapest rooms on the inside of the stadium. They didn't even have windows to the outside, so it was kind of dark and dank and mildewy. There were three of us—Bobby Dreher, I guess it was. Billy Drake was another one. Billy is dead. He was a Grumman test pilot who got killed out at Edwards Air Force Base flying a new Grumman airplane. Bobby works up at Langley. They had trouble trying to keep up with us because they couldn't shoot snooker during final exam week. They had to study. I could never do anything like boning up. If I didn't know it before I went in there, I wasn't going absorb it in a two or three night period. I've been that way all my life. I was fortunate that I was born with two great assets. I was born lazy and with no ambition. *(laughs)*

ROBBIE: You sure went a long way for somebody who is lazy and with no ambition!

GUY: Because I was lazy, I became an outstanding supervisor because I didn't mind letting other people do the work. I didn't have to do the work, or try to be smarter than the sum of the guys in my division. But I knew enough to guide them. And because I had no ambition, I was never a threat to any boss I had. I never walked over any of them. I walked *around* quite a few of them, but I never walked over any of them. And I never worried who got credit for the work. I think I parlayed those two things other people think are not too hot into a pretty good career. And I had an awful lot of fun doing it. I don't think anybody could have worked in anything more exciting than Max and I and Paul did. We were right in the forefront of everything. We always had the best tools. We had tremendous support in everything we did. You just couldn't ask for a better deal.

KEN: Paul went to LSU also. Did you know him there?

GUY: I didn't know Paul that well. I knew *of* him at school, but I didn't know him. He graduated two or three years ahead of us. Because he was working at Langley, he had an exemption from going into the service because he was working on defense related research. So, Paul didn't go into the service. By the time we had gotten to Langley, Paul had been working there for six years and had kind of gone up the ladder. Gilruth had brought him in—Paul had headed up the small wind tunnel there at one time. He also headed up the fiscal office, the payroll office at

Langley one time. We got Paul's name from the head guy who taught Max aero-engineering.

The guy who headed up the aeronautical department at LSU was an old German from World War I, Ernst (Fritz) Maser. He designed some real fancy airplanes back in the '30s. The fastest airplanes in the world had been designed by this guy. He had a very small organization. He never had any more than ten or twelve students. He handpicked all of his students, so he knew they were all topnotch people before he accepted them. He knew they would get through.

ROBBIE: And he was your professor?

GUY: No, he was Max's. I knew him and I worked with him. I got to know him fairly well back in the early '40s. Fritz was head of a Navy and Air Corp Cadet training program at LSU. As I already had my commission, but had not been called to active duty because I was being allowed to complete my education, I was hired at the enormous salary of $5.00 an hour to give these cadets military training. That's really how and when I got to know Fritz.

ROBBIE: So you drive up there in Max's father's car, and say "Hey Paul, we're here?"

GUY: No, he didn't know we were coming up. We got there and the first person we met was Adeline. She was the receptionist in this fancy building. There was a rotunda and it had paintings of the history of flight on the large hemispherical dome. There was this big circular table and she sat there where you went in. She took one look at us and kind of turned her nose up for some reason—I guess because we looked like two bums. We were wearing very loud Hawaiian sport shirts and either khaki or navy gray work pants. We had opened-toed sandals with no socks on. I think we slept in our clothes. We weren't anything special looking. (Later on, she must have changed her mind because she asked us to give her a ride home after work.)

But they didn't give a hoot—they didn't care about that. They were totally informal. The whole operation was informal. Paul never wore a tie, in fact. He suffered with the heat so always in the hot summer he had a big towel wrapped around his neck. Whenever we had any distinguished visitors, he used to keep an old seersucker coat hanging on a tree in his office. This office was just one big office with everybody in there. Paul had a desk at the head of the room. He'd go grab some kid who had a tie and say "Hey, I want to use your tie" and take the tie off him and put it on and go to the meeting if he thought he had to be dressed up. *(laughter)*

ROBBIE: When Paul saw you guys, what did he say?

GUY: He was *enthused* because here are two young fellows who had engineering degrees looking for a job. They had hardly hired anybody with degrees. Most of the people had gone off to war. It wasn't that easy to get good people. He made us an offer right on the spot.

ROBBIE: What did he offer you?

GUY: A job working in the Pilotless Aircraft Research Division with PARD founder Bob Gilruth. They hired me to work on rockets and they hired Max to work on a type of propulsion called ramjet. Paul said to me, "You are a chemical engineer. We need someone who can work on things called liquid rockets." Well, once I got up there, I found out that liquid rockets were not what we were supposed to use—solid rockets were so much easier and cheaper, and we could handle those so much better.

ROBBIE: Is that what got you interested in propulsion?

GUY: Yes, I got interested. I taught myself with a bunch of books from the library. I set up test facilities and later on, I built myself an experimental solid rocket plant so I could try out some of my own ideas on high performance solid rockets that I had patents on. They had a bunch of surplus military rockets we were working with—little teeny things between two and five inches in diameter. I got interested in them and a lot of other things. I was part of a service organization—I provided a service to all the people there. I was not in the forefront like the other people. I kind of worked in the background.

I found out a lot of things—that if they would let me make changes in the rocket designs, fix the rockets up and do things a little bit differently, I could help enhance quite a lot of the aerodynamic data we got. So I got in the business of cutting rockets in half, building bigger ones, gluing two of them together, putting them in a lathe and cutting them down, doing all sorts of things to change their performance characteristics, which would help the engineers get better information. We developed a bunch of little techniques.

One of Bob Gilruth's real strong points was that he believed in absolute simplicity. He believed in using your head rather than the machine to do things. He would say, "Use your head!" I learned a very important lesson which I began to put together from him—that passive systems are much more difficult to design than active systems. Passive systems are where you press the button and everything happens according to the laws of physics and chemistry without any intrusion of anything—a valve that you have to adjust, something that has to turn something else off or on—he was real great for that. He never let you use a lot of complicated devices.

ROBBIE: Because there would be more stuff to break?

GUY: Yes. They cost money and they can go wrong. They can fail. So what you do is use the laws of physics and chemistry and put those to work for you so that everything happens the way it's supposed to happen on the way up. With the basic design, you design that into the system so that nothing active had to come in and intrude on it.

I caught onto that real quick, and that's why Gilruth and I got along real well. One time he had us study the dynamic qualities of an airplane to make the airplane stable. If you pulled back on the stick and then you release it, then your airplane goes through oscillations, and those oscillations tell you how stable your airplane is. He didn't want anything to go bad, so I said, "I can fix you up. I can put little rockets about so big—about a half inch in diameter and four inches long—that fire at right angles to the model, and that's going to knock the nose down and make the thing oscillate. I can set those so they go off in one second intervals using delay fuses. I'll put eight or ten of those and I'll fire them all from the ground and they'll all go off at different intervals, so as the airplane is slowing down, you get the entire speed range from supersonic through subsonic. We started doing things like that.

There were a lot of other little things that people wanted. One of the guys said there was an aerodynamic quality they wanted to measure called "damping in roll." He came and talked to me about it. He said, "I have to have something to make the airplane spin, but I can't use the rudder to make it spin." He said, "If I had something to spin it," and I said, "Then I'll design you a rocket to spin it, and I can tell you how much force the rocket's using to spin it—will that help?" and he said "Sure." So I designed those things—a lot of little things like that.

One of the guys in the wind tunnel called me one day and said, "Hey, we'd like to study something in the spin tunnel, but we can't. There are a lot of things we can't do." The spin tunnel is a vertical tunnel—the air goes up. You throw a model in there and then you put it into a spin and you blow the air on it, enough so it keeps it floating there, you take pictures of it as it floats around as it spins. He said "Well, you really can't tell very much about the aerodynamic forces on there or the other forces, because they are all in the wash. It spins around, it's nothing but the air that is displaced that it's spinning back into. I can't tell what forces and what direction you need to apply to cause the airplane to recover from a spin" But he said, "If I had a way that I knew what the forces are that cause the recovery, I might understand what I have to redesign in the airplane to make it recover from a spin." And I said, "Well I'll design you a little rocket that can provide a known force and moment you can put in there." And I built little rockets that fire a couple of seconds. They could fire those at all the various axes and cause it to de-spin. Then they could study the spinning characteristics, the forces and the inertia that it went through in the tunnel. That was a big help.

ROBBIE: You built rockets that were tiny?

GUY: They were about a half-inch in diameter and an inch long, and produced three ounces of thrust for two seconds. They were mounted on a little model airplane in the tunnel and what they would have inside was a little switch. The tunnel had a large solenoid coil in its throat. When you energize that magnetic coil, you close this little switch to fire those little rockets without any external wires into it. There was a small battery and initiator inside the model.

We did all sorts of things. It was a kind of free-swinging outfit. When we tested airplanes in the wind tunnel, you had these turbojets hanging on there, but there was no exhaust coming out. With no exhaust coming out, it doesn't give you true conditions. The exhaust acts like a big solid body. It's like trying to stick your finger into a hose. The jet exhaust pushes the free stream back out. John Stack asked my advice on installing something to simulate turbojet exhausts, so I suggested a hydrogen peroxide mono-propellant system, which would give about the same exhaust characteristics as a turbojet.

I'll tell you a little story about my rocket plant. You see, Lewis was the propulsion research center and Langley was supposed to be in aerodynamics.

ROBBIE: What were you doing at Langley instead of Lewis if you were in propulsion?

GUY: I was supporting activities at Wallops Island in Virginia and they used rockets all the time. I wanted to build this rocket plant. I had some ideas about a brand new spherical rocket that I thought would have some advantages. I wanted to be able to build one to prove it. So I needed to build a rocket plant. Headquarters would absolutely not approve of building a rocket plant at Langley field. They said, Oh no, only Lewis could have a rocket plant.

Bob Gilruth came to me and said, "My signature authority is worth $999.99. If it's over $1000, you'll have to go to headquarters for approval. Can you buy every piece of equipment you need for that rocket plant for under $1000 per piece?" I said "I think I can." He said, "Don't worry about the building. We'll call that a model assembly shop or something. We'll build you the building. That's no problem." I had to take some of the low bidders for equipment I needed. But, I built me a rocket plant. I bootlegged it! The next year, because of what I did with that plant, headquarters decided to give $225, 000 to build me a good one, a bigger one.

Development of the American Space Program: Early History

GUY: The space program really got started back in 1957 with NACA—the National Advisory Committee for Aeronautics. I worked for NACA in PARD, and what we did was aeronautical research using free-flying rocket-powered models, and that research had a great influence on the ballistic missile program, including the shift from big liquids to all solid.

With these small launch vehicles, we were getting up into the ionosphere, and at the same time getting a velocity that was short of orbital velocity. Just about the time Sputnik flew, in October of 1957, the Air Force had a program eventually called Dynasoar. It was either a winged bomber or a winged orbiting space surveillance aircraft. There were two versions of it—one with a semi-global range—it could fly halfway around the world. Then they discovered that with available propulsion technology, you could actually go into orbit and fly it *all* the way around. The Air Force wanted to develop this winged space bomber to be able to fly and bomb anywhere in the world.

That program was the outgrowth of something happening in Germany during World War II. The Germans had developed the V2s, and were looking at the very large V2 they called A4 and A10. Its purpose was to be able to bomb New York from Berlin. The way Dynasoar came into being was that the Army had gotten all the German scientists at the end of World War II. They had gotten von Braun because the Air Force did not exist at that time. The Air Force, once it came into being, was kind of miffed because the Army had gotten the group who were in the business of developing rockets that could bomb cities, in direct competition with bomber airplanes.

Since the Army had von Braun, the Air Force decided on a little one-upsmanship. So they imported von Braun's boss, one of the German generals in Hitler's high command. His name was Walter Dornberger. He's the one who wrote the book called *V2*, about how the V2 developed in Germany. The Air Force set him up at Bell Aircraft Systems in Buffalo, New York.

ROBBIE: What year was this?

GUY: This must have been in the early 50s. He was Chief of Research, Chief Scientist at Bell Aircraft. The Air Force had some studies which were conducted under the code names Robo and Brass Bell. They were top-secret studies. They were basically on glide bombers, not like Dynasoar. These studies eventually grew into a proposal to build some of these things.

In October 1957, NACA was asked by the Air Force to convene a group of people and to take a look at this program they later called Dynasoar, and to advise them of the feasibility of doing this program, possibly as a continuation of the high speed flight research program past the X-15. People from throughout the agency were invited to NACA's Ames Research Center at Moffett Field, California to sit in on these discussions. It was the week when Sputnik was launched.

The U.S. had its own satellite launching program, called Vanguard, which was the responsibility of the Naval Research Laboratory. That grew out of a program called Viking, which was a relatively large liquid-fueled sounding rocket that could probe the upper atmosphere and gather all the properties of the atmosphere—density, temperature, radiation and various other things that scientists were interested in measuring. This program wasn't doing that well at the time. It was having many failures and was way behind schedule.

There was a young X-15 test pilot named Neil Armstrong who was there too. We all flew out to California to Ames Research Laboratory near Mountain View. It's in the San Francisco area. We sat around for about three or four days and had a number of technical discussions about what the Dynasoar should do and what it should look like.

Another fellow named Hartley Soule from Langley was also there. Hartley was the guy who started the Flight Research Program at Edwards Air Force Base. The last big one they had was the X-15 Mach 6 research airplane.

We looked at Dynasoar and saw that there were a tremendous number of obstacles, but none of them were insurmountable. They were strictly engineering problems that somehow or another, if you worked hard enough and you did enough research in some specific areas, could be overcome. Nothing seemed to violate any laws of science as we understood them.

Ultimately, we have proven that the shuttle is nothing but a very large Dynasoar. When you really look at it, the Dynasoar had a similar shape. We didn't have very large launch vehicles at the time. And the shuttle basically had wings just about like the Dynasoar. It glides back in just like the Dynasoar was supposed to do. You see, the Dynasoar was feasible—it's just that there's a lot of what I call *collateral technology* that has to be developed. You want to do things, but in order to do them something else has to happen to allow you to do those things. You have to develop the materials and understand the aerodynamic forces, heating, propulsion, structures, materials and guidance and control. And all those things you need have to be small and lightweight and have to use little power and last a long time unattended. From 1950 to 1970, when we were working the space program on Mercury, Gemini, and Apollo, we were developing all that technology that would allow Dynasoar to happen.

We had a few things we were interested in doing. I was interested in building an all solid-propellant launch vehicle. You see, most of the launch vehicles are liquid-propellant, except that suddenly they shifted over from liquids into solids for the ballistic missile launch vehicles, for a lot of very good reasons. Many of the reasons were the results we got from launching these multi-stage solid – propellant rocket vehicles at Wallops Island. We demonstrated a very high degree of reliability and very simple operation. Their simplicity, lack of much ground support equipment, and instant readiness makes them much easier to do than the big liquid system.

Max had gotten interested in putting a man in space as a result of discussions with Al Eggers and in addition to looking at Dynasoar. And it looked much easier to do than Dynasoar. The easy way looked a lot like this ballistic missile nose cone technology. We would build a little blunt nose cone and it didn't have to have all the controls or everything on it. It was going to be a very, very simple thing to do compared with doing a great big thing like Dynasoar, and really a way to get man into space in a hurry.

So out there at Ames, Max and I cornered Dr. Dryden in the lobby and we were doing a little bit of lobbying ourselves, telling him that we were not going anywhere fast and if he would just give us a chance, we could develop this little launch vehicle—which later on became a four stage solid propellant rocket satellite launch vehicle called Scout. (The Scout had a very successful history. It's no longer flown but it was flown up to about two years ago before the program was cancelled. During its lifetime it had the most reliable launch record of any launch vehicle.) And Max said we could put a man in orbit if we were given the authorization to do so.

Dryden said that Eisenhower had told him that the Naval Research Laboratory was the only one that was going to put a satellite into orbit. Later it turned out that he had to recant, because the Navy got so far behind that von Braun launched the first little satellite with the Redstone and two stages of solid propellant rockets. And then Max talked to Dryden about putting a man in space. Dryden didn't take too kindly to that. I think his comments were, "Shooting a man into space is like shooting a person out of a cannon"—or something like that! And frankly, Al Shepard's flight was about like that. I think when you really get honest about it, it was more about PR than anything else.

So Dryden said NO we are not going to do any of that. We got back to Langley and talked to Bob Gilruth, and Bob said, "Well, you guys go ahead and work on it." He said, "We won't tell anybody." Most of the greatest work we ever did is what we called "bootlegged." We got it started before we ever had authorization to do it. We would develop the concept, prove that it worked, and after we proved that it worked by bootlegging it, why then they'd give us authorization to do it. We always did think like that back in the old NACA days.

The National Advisory Committee for Aeronautics (NACA)

GUY: The NACA was a very unique organization. We probably had somewhere around 7 to 8 thousand people in the organization—scientists, administrative people, and crafts people. We had a lot of very skilled people who could make anything you wanted. The machine shops—they could make anything that anybody could conceive of. We had instrument machine shops that could make instruments. We were completely self-sufficient—our organization—pretty much. We did not have to do a lot of outside contracting, except for maybe the large construction. To manage that organization of 7000, the total headquarters/professional staff was 58 people in Washington! The proportions are quite a bit different these days. But the NACA was not a political organization. It wasn't big enough that anyone would want to grab it. It really didn't have to be defended, and there were a lot of people who supported it because it did nice work.

The NACA was a funny outfit. I worked there for 17 years and I don't know anybody who ever got an award from the organization. The way you were awarded is you got to work on the best jobs, you got ideas heard, and you got promoted. They didn't give you a piece of paper or a plaque. Everybody knew who was doing the job and what would happen. Then, if you did something good, everybody in the organization who was good wanted to work with you on your team or on your ideas. It was "the cream rose to the top." That's the way the organization worked, and everybody understood that. No one had to worry about anybody patting them on the back because they knew what they did and everybody else knew what was done. Occasionally somebody got a big award from some place outside the agency. A modern parallel to the way the NACA operated is what happened in the computer industry in Silicon Valley in its early years.

Caldwell was another one of the wonderful people in the outfit. He designed spacecraft. He said his neighbor always asked him, "Well, I see all these military people getting all these outstanding ratings and superior performance awards—how come I never see any of you guys ever get one?" Caldwell said, "Hell, you got to be superior just to work for the NACA!"

Back in March of 1958, someone in the Eisenhower administration apparently decided they wanted a space agency. The troops in the trenches didn't really know that much about it. But I think the NACA headquarters was told that they were going to become the nucleus of the space program. Most of that is a result of the work that we had done at Wallops Island with Bob Gilruth and his Pilotless Aircraft Research Division--that's why we became a space program--the aeronautics part of course was fairly well established throughout the rest of the agency.

Formation of the NACA Space Committee

GUY: Gus Crowley, who was the deputy under Dryden, formed a little committee in the middle of March, 1958. Its job was to take a look at the NACA's role in space.

ROBBIE: Where did you meet?

GUY: I think we met the first time at Langley about fifty years ago. There were many, many meetings after that. This grew into something of a permanent affair. And not all the people from the original group stayed with the group. There were various reasons why a lot of them didn't continue, and various reasons why there were substitutions.

Robbie: Was this the Space Task Group?

Guy: No, this was a year before the Task group. This was just a group of people mostly from Lewis and Langley thought to be leaders in their fields who they thought could come up with the program and get it funded.

Max was very active on this committee, but I don't know if Paul was that active or not. Paul was kind of Gilruth's Chief Executive Officer. He did everything Gilruth didn't know how to do or didn't want to do. Gilruth had some things that he was super at that he loved to do, but like all of us, we can't do everything, and he knew who to choose to do all those things that needed to be done that he wasn't going to mess with. Paul was that type of person—in addition to being very good in his own field as a scientist, he was good at all kinds of planning and administrative stuff.

Bill O'Sullivan was our resident egghead. They call them "nerds" these days. He was a kind of science type. He's the one who came up with this great big balloon, this hundred-foot diameter balloon we launched in space, called Echo. You could see it from everywhere. It was a real bright star and we had two of them up at once.

Robbie: What did they do?

Guy: Well, you could monitor radar signals—radio signals actually—sort of like a passive communication satellite, except satellites now act so you can send the signals to them, amplify them, and retransmit them. It was something that everyone could see in space. It was a lot of good PR as well as for measuring the density of the atmosphere, because the number of molecules up there is what slows the thing down so much. It was measuring the decay rate. It was huge and it weighed practically nothing, so it had a very measurable rate at which it could come back into the earth. There were a lot of things you could do with it.

ROBBIE: *And* it was good PR?

GUY: It was big! Everyone could see it and you could bounce a radio signal off it—well, after the Russians launched Sputnik, someone drew this a cartoon that showed Sputnik and our balloon like two children in baby buggies, and Eisenhower says to the Russian guy, "Ours can talk!" I could still pinpoint the exact day, the first time I came down here to Texas. I had eaten a barbecue dinner over at a friend's house, because that night at dusk I was watching, and both of these two balloons crossed right overhead. There was only one time over that place that that could happen. I could go back and track down the date and hour.

Well, anyway, basically the task of the members of this committee was to find out what was going on in the military and the Department of Defense that was space-related, and see if those were the items that we wanted to budget. We caught an airplane from Langley on Monday morning, we went up to Washington beginning in April, almost every week through October, and we spent four days a week in Washington.

ROBBIE: What would you call the group?

GUY: I didn't call it anything. I was just going to NACA headquarters to work on the space activity. If it had a name, I'm not aware of it. We were not a very formal organization. The other thing is that I don't think you'll find any minutes to the meetings. We verbally reported to our leaders until later on when we came up with the budgets. Then all our input was fed into Ray Zavasky, who pulled the whole budget together and published a little booklet for Congress. All I know is we were trying to pull together a space program for the NACA.

Basically, we were cloistered. Nobody in the NACA headquarters except for Crowley, Dryden, and maybe Ira Abbott, who was one of the three senior people up there, and Clothaire Wood actually knew what was going on, and we did not speak to anybody else. We had a big conference room up on the top floor. It might have been Dryden's executive conference room or something.

Robbie: Is this still in 1958?

Guy: Yes, this is 1958, from March through October. And usually we'd spend one day back at Langley trying to keep the store. Max was a branch head, which is a pretty important job. Paul Purser was a branch head, and I had two sections in Paul Purser's branch, which is unusual, but we couldn't find anybody to do it, so I headed up two sections—the Rocket section and the Materials Research section in Purser's High Temperature Branch. We were doing research on all the re-entry stuff, which turned out to be very valuable, all the protection systems and materials, things like that. We were working on that.

NACA headquarters was on 17th and F, right around the corner from Lafayette Square, by the White House, in fact. The Madison House became part of the NACA headquarters. We were supposed to find out everything DOD was doing in space and bring that into NASA, the new organization. It was going to be the exclusive organization to work on space activity, but we didn't know it at that time.

Robbie: So in this little group, you were ultimately talking about founding NASA.

Guy: We didn't know it! Some people knew we were talking about founding NASA. Dryden knew it, I think, and Crowley knew it, but no one told us when we went up there that we were going to become NASA. The fellow who wrote the Space Act was named Paul Dembling. He was NACA's General Counsel and the only attorney in the entire NACA practicing law.

Anyway, we had all the top security clearances that you needed to get into the meetings going on in the Pentagon. Charlie Zimmermann would give us lists of meetings going on in the Pentagon and the Navy department and over there at the Atomic Energy Commission. Somehow the word got to these organizations that we would be permitted to attend to these meetings to listen, be quiet, or participate in the meeting.

We'd go to all these meetings over in the Pentagon. I was interested in propulsion. That was my particular field, since I headed up all the rocket work at Langley as well as the high temperature materials work. I got into high temperature materials research because rocket exhausts had the highest temperature of anything around at the time. So the earliest research work we did was using rocket jets with the materials and getting very high heat fluxes so you could test the high temperature materials. We'd go to the Pentagon, the Navy department, or wherever these meetings were, and attend these meetings and find out what was going on and see what we wanted to include in the budget we were designing—what we thought we should budget for some of these items because the military was not going to be permitted to continue this work in these areas necessarily. Not as a space program per se.

We were to come up with a budget as though the NACA was going to be a space agency and was going to be active in space. It didn't say we were going to be the *exclusive* thing in space.

Creating the Space Program: From NACA to NASA

ROBBIE: So you have this group and your meetings go on through October and you come up with a budget.

GUY: The way we reported the facts is that every night we would walk back through Lafayette Square. There was a liquor store on the way home, and Bob Gilruth and Abe Silverstein always had their room on the first floor at the Francis Scott Key Hotel, next to the coffee shop. We sat around and had a few drinks and we discussed what the daily activities were. There were no minutes at the meetings. That's the way we communicated, at the hotel. The young fellows would tell them what our opinions were about everything we'd gone through that day, whether we had been over at the Pentagon or out to the AEC and what programs we thought were good and bad. We would get some feedback. We did that almost every night.

There was a time by which we knew we were going to become the space agency. The big question was, who decides the NACA would become the space agency? What other organizations are to be included? There were no decisions made as yet as to what organization would comprise the space agency. One organization was the Naval Research Laboratory group that did the Vanguard and the Viking, and did a lot of upper atmosphere research and science—

would that become part of it? One of the decisions we had to make was, we were we going to build another Center in the Washington area, and if so, where is that going to be?

The Army, the Air Force, and Wernher von Braun

Guy: Dryden got wind that there was a big power struggle between the Air Force and the Army on account of the ballistic missiles. There were two competing programs, the Air Force Thor program and the Army's Jupiter program. They were both about the same size and intended to do the same thing.

The US Air Force was originally part of the Army—the Army Air Corps. When it became a separate branch of the service, at first it didn't have any organization. It didn't have any laboratories. The Army developed all of its stuff in-house through the use of arsenals. They developed tanks, and they developed guns, and the Navy also had its organization that developed things. The Air Force didn't have anything. But the Air Force had some smarts. They understood that a big bureaucratic organization cannot lobby for itself. The Air Force didn't have any of this stuff, so they decided to get into bed with American industry. When Eisenhower left office, he was very concerned about that sort of thing.

But that's the way the Air Force came into being. They gave contracts to industrial organizations and said, "You scratch my back, I'll scratch yours." The Air Force had all the industry and all the money in industry to lobby for the Air Force. The Army didn't have any of that. In the power struggle, the Army lost out on this big long-range ballistic missile. They were confined to what they called tactical missiles with a range of 500 miles or less.

But here you have this big organization down in Huntsville with von Braun building these Jupiters and Redstone ballistic missiles for the Army. The Defense Department didn't know what to do with it. Von Braun was one of the biggest proponents of space I've ever seen. He was a master at public relations. He understood that Big Bureaucracy can't be heroes in the public's eyes. Only people can be heroes. He was the guy who took credit for everything that went on down in that organization. Wherever he went, when he stepped off the airplane, people wanted to talk to him. He *was* the organization. That's the way he played it, and that's how he kept on getting budgets and how he got Congress to listen to him—by becoming an international hero. Very few knew how to deal with that, but he understood that.

ROBBIE: Did people that work in his organization who had actually done some of the work he was taking credit for, did they resent that or did they understand it?

GUY: No, that was all part of the deal. They understood that, I'm pretty sure. He was always on what I called the "borscht circuit," giving speeches at lots of rubber chicken, mashed potatoes and peas dinners. He was out selling space and missiles. He was a super salesman. He was selling himself, basically, but selling the program with himself. In fact, no one even knew who his deputy was. I could have talked to 95 percent of the people in the industry and ask them who von Braun's deputy is and I would say 95 percent even of key people couldn't tell me what the man's name was. His name was Eberhard Rees and he kept the store when von Braun was on the borscht circuit. He was a very mild, meek little guy, at least outwardly.

When Dryden got word that he was going to have to take in von Braun's organization, which was the part of the Army Ballistic Missile

Agency that later became Marshall Space Center, he asked me to go down there to Huntsville and case the joint. That was probably in July or August of 1958. At the time I was a GS-15—that's the equivalent to a colonel in the Army, so it shook them up that I was the only person on the airplane. I had my own private airplane with a GS-15 pilot flying me into the Redstone Arsenal airport. That kind of shook them up a little bit—protocol, you know. They were supposed to have at least a bird colonel meet me at the airport and show me around.

I went down there and found out how many people the Army was going to keep. I didn't ask about that—the way I went about talking to people was to find out what was pretty much part of the Army and what was part of the Redstone or Jupiter and the other part of the program that von Braun was involved in. The Army had small launch vehicles and small tactical weapons systems and anti-aircraft missiles. They had two separate organizations going. Von Braun headed one and somebody else headed the other.

ROBBIE: And they were going to split because the Army was going to move out of space?

GUY: They were going to split. The Army didn't know they were going to be told to split, but they were distinct and separate in a way that you could divide them pretty easily. Maybe some people were a cross between both and they could settle that later on. I think I found out they had somewhere between five and six thousand people in von Braun's operation. Boy, Dryden was real excited because this thing he was going to have to take on was almost as big as his whole outfit. I don't know how many we picked up, probably around 5,000 people we picked up when they created the Marshall Space Flight Center. It was a very interesting time.

I came back from Huntsville and gave Dryden a verbal report on that, and he was surprised he was going to have to take on so many people because that operation was two-thirds as big as the whole NACA was at the time.

Later on, by the time October came along, we all had our input in as to what we wanted to do. We wanted to start Scout, we wanted to start Mercury, and I wanted to start large solid rocket programs. There were a number of liquid rocket programs that were ongoing that the military had that we were going to pick up.

Formation of the Space Task Group

GUY: Our committee started meeting in DC in March of 1958. We became the space agency on October 1, 1958, when Congress authorized the creation of NASA. We had been bootlegging the space program from October of 1957 until 1958. We were doing the in-house study—getting prepared, finding out what it would take for us to accomplish all these tasks we wanted to do, how we were going to go about it, what the thing was going to look like, and how tough it might be to do it. We all worked on various aspects of those programs--the Scout and the Mercury program were the big things we did right in our own organization. Various other people worked with the people on space science and application.

ROBBIE: Now tell me how the Space Task Group was formed.

GUY: They needed an organization to run the Mercury program. They weren't going to let Langley run the space program, because it was more like a research outfit. They also created Goddard then. Some of the folks involved in this exercise and other people they brought in actually had looked for an area to build Goddard. There was a farming cooperative called Greenbelt Cooperative. They had some farmland up there between Washington and Baltimore called Greenbelt, Maryland. That's was chosen as the site for the Goddard Center.

ROBBIE: What was Goddard's purpose?

GUY: Space science. The Space Task Group was originally to be assigned up there. The person who created the Space Task Group was the director of Langley. He didn't need anybody's permission—he just wrote a memorandum to create the Space Task Group and signed it "Floyd Thompson."

ROBBIE: Why was the Space Task Group needed? You guys already had your jobs at Langley.

GUY: It needed to be identified as an organization that was going to manage this very large manned space program distinct from Langley. Headquarters was going to be involved in a lot of things. Langley had never been involved in public affairs, the whole bit. At Langley we didn't have any Public Affairs office—we didn't need one. The work we did, we thought spoke for itself. We didn't have astronauts. But there was going to be this big hullabaloo about this space program so they needed a separate organization.

And Thompson was very magnanimous. He vacated buildings for the Space Task Group and gave them space to house their offices over there and assigned people to the STG. He had a cute little way of doing things. Thompson said, "Bob, you can have a lot of people but what we are going to do is for every one you take, I'm going to give you one." So you have a balanced organization, I guess, was the inference—you are not going to grab up all the good people. This was also the policy between NACA Headquarters and the Centers. Whenever Headquarters wanted to pick some bright youngster and offer him a job in Washington, the next one they needed was the Center's choice.

ROBBIE: "Take the ones you want but I'll give some I don't want."

GUY: "For everyone you take, I'll get to give you one." It's like choosing sides in a kid's baseball team.

ROBBIE: And did Gilruth actually do that?

GUY: Yes. All the people who came down to Johnson Space Center were pretty proud of coming down here, but not all of them knew which group they were in—the ones Gilruth picked or the ones Thompson made him take.

ROBBIE: They didn't know whether they were chosen or sent!

GUY: No. Most of them were very good but some of them were dull and not very bright.

The Leaders: Bob Gilruth, Abe Silverstein, and Joe Shea

ROBBIE: What was Bob Gilruth good at?

GUY: People. He was one of the greatest people persons you ever saw. He knew who to trust and who not to trust, who to appoint to key positions and how to let them do what they were supposed to do. He knew how to communicate with you. I worked with the man for most of my professional life, and he'd call me and we could discuss things. I had things that I wanted to talk to him about, and I'd say, "I'd like to come down." He was never formal. When I first met him he used to come around to see me, and we weren't very fancy. We had old gray desks and no curtains, no rugs on the floor. Everyone had a wastebasket. He made a point almost every week to come by and talk to everybody in the organization. He always sat like he was sitting on a pot on my wastebasket and talked with me.

ROBBIE: He'd sit on your wastebasket without turning it over?

GUY: Without turning it over. It usually had trash in it, and he'd turn it over if there was no trash in it. He never once told me what I should do. I never once left not knowing what I should do. I don't know how he did that. I knew whatever I did, if I chose to do it, when I walked out of there I had his full, unmitigated support. I don't how he did that with everyone, but most of us, I think, kind of felt that way. He understood man-machine relationships, which made him great on the space program. He knew what the role of man was and how you were supposed to make his job easy, and what had to be done in order for a man to fly. He had a gut feeling about all those types of things. He was great at that. But I think his greatest asset was gathering the right people and knowing who to trust and who not to trust.

He was an absolute gentleman, but I could always tell when he was angry with someone. If he began to squirm during a presentation, I knew the presenter was in trouble. There's not that many people who have that knack, but those of us who worked very close to him always knew when someone was in trouble. His favorite expression when things didn't work out just right was, "Well, I just don't think we had enough talent on that job." And that's about all he'd ever say. That was the way he dealt with things. He never demeaned anybody. He was just a real gentleman, a really great man.

Abe Silverstein was different. He was a very domineering person, very bright guy too, very talented guy. Abe was an aircraft propulsion type, turbo jet and other aircraft engines. He became Director of the Lewis Research Center, which is in Cleveland. He was very gruff and he tried to put the fear of God in everybody. Most of the people who worked for him really were afraid of him. Most of the people who worked for Bob Gilruth loved him. None of us had any particular fear of him because he was such a gentleman in dealing with us.

And I learned quite a few things from Abe, I guess, like the fact that he could be buffaloed too. Like the time I went up to rescue the Scout from being transferred to Washington, I had him right where I wanted him. They were going to transfer the Space Task Group to Goddard. Abe wanted to put the Space Task Group where he could have it closer to Washington and have a little more control over it. He also wanted to transfer the Scout up there; the Scout was something we were doing at Langley. Abe had a committee that said "There ain't no way on earth that we would ever fly that thing." It was too long and limber, and we didn't know enough about thrust vector control systems and about thrust misalignment and solid rockets, and the thing would never fly. We didn't pay much attention to them because we knew better.

So Abe told Floyd Thompson he was going to move it up there. At the time, we had budgeted about $27 million dollars for the entire program. There was also a lot of talk about the fact that the Navy had some obsolete Polaris missiles there, submarine-launched guided missiles. Someone was trying to force us to use those old Polarises to make a version of the Scout instead of using the Scout itself—a very expensive move.

We had some meetings at Langley and I finally convinced Floyd Thompson that we weren't going to do anything like that. Thompson said. "Well, if you want to do this program, you have to get the money." We needed about $8 million dollars to get this program started, so Thompson said to me, "You go up there and talk to Abe and you get the money." So I went up there and talked to Abe and that was another interesting thing. By this time, there were a bunch of people at headquarters from the Army, Navy, and Air Force who transferred to NASA to get a grade raise. They wanted me to brief them and then they would go tell Abe about what was going on. I said, "No, we ain't gonna work things that way. I'm up here and I'm gonna talk to Abe." They said, "You're only a GS-15 and we are much higher in the organization than you." And I said, "Well I don't care about that—I'm the Center Director as far as you're concerned. I represent Floyd Thompson and Thompson didn't tell me I had to talk to any of you people. I'm going to talk to Abe. And if you want to come, you'll have to get Abe to invite you."

That's the only time I've ever seen Abe Silverstein ranting and raving. He said, "I told Floyd I'm going to transfer that thing up to Goddard." I said, "What size boxes do you want them in?" He said, "What do you mean, boxes?" I said, "We don't have many people assigned to work programs full time. They are spread out all over. None of us want to go to Goddard, so if you want the program, just tell me what size boxes you want us to put the documents in." And oh, he started ranting and raving and jumping up and down. That's the only time I've ever seen him flustered.

Afterwards he got calm, and he said, "All right, you go back and you tell Floyd I'll send the money. But there's one thing you'll have to do before I send the money down. You are going to have to put a project office together and assign people to it and I'm going to have to know who is working on it." So that was one of the agreements we had. And I went back and we created the Scout Project Office and that's what ran the program. That's how we kept the Scout back at Langley.

KEN: Oh, what a story!

GUY: That's when Bill Stoney got to be named head of the program—the first Scout Project Manager. That was the one time I had Abe where I wanted him, and I knew it. When I reminded him of that he said, "Oh, no, I would never do anything like that." I said, "How come, Abe?" He said, "I would never take the baby away from its mother."

KEN: Joe Shea was program manager for the early Apollo. How did Max get along with Joe?

GUY: Oh, fine. You heard about the big foot race? Max used to jog every morning, run a mile every morning. Joe Shea ran too. So, they went to La Porte and they had a big foot race over there. They ran a mile. I asked Max, "How fast are you going to run?" He said, "Well, I'm going to run under six minutes." I put a ten dollar bill on the table. I said, "If you run under six minutes, that's yours. If it ain't, you're gonna match it." They went up there and they ran the race. When they came back, I asked Max, "How did you do?" He said, "Well, I didn't do too bad. I came in second. Joe Shea came in second to last!" *(Laughter)* To be fair, Max spotted Joe a few years.

There was another race. Pete Conrad had told Max he could run a mile in SkyLab. Max bet him he couldn't. Remember, in the Skylab movies you'd see Pete running around the ring up there? Well, Pete ran around the ring for a mile. He called down and said, "Tell Max I ran my mile!"

End of an Era: Langley after the Space Task Group

ROBBIE: You were not part of the Space Task Group?

GUY: No, I didn't have to be because I supported the Space Task Group while I was at Langley until they moved down to JSC. I was available to them for any of my particular specialties. They had what they called "Capsule Coordination Committees" up at McDonnell Aircraft, talking about all of the various aspects of the Mercury program, and I was one of the members of a Capsule Coordination Committee. I flew up to St. Louis quite a lot during that time frame. Even though I was not part of the Space Task Group, I was available to them anytime they needed any expertise I had.

Then, after the Space Task Group was formed, they came to JSC in 1962. I was still back at Langley. I had a suspicion there that they were stealing so many people from our outfit that Thompson decided he needed to keep some continuity of experience, and so he wasn't going to let Gilruth have me, even though I had worked on the program—I'm not sure. But I got a big reward out of that—I was one of the very few branch heads that had what they called "an excepted position." It might be equivalent to the Senior Executive Service. Right now it would be above the GS-15 level. Not too many at that time had that position.

ROBBIE: So in other words, you got promoted at Langley?

GUY: I think that was for me not raising a lot of ruckus wanting to leave.

ROBBIE: Did you want to go with the Space Task Group?

GUY: I suppose I wanted to go since it was all my old cohorts. There were some good people left. The real drivers were pulled into the Space Task Group—most of the people that Gilruth picked. He knew who he wanted.

ROBBIE: What were you promoted to?

GUY: I got Paul Purser's old job. I became head of the High Temperature Branch. There were a lot of branch heads who didn't have the same grade I did. I had the equivalent of two or three grades higher than most branch heads.

ROBBIE: That gave you, along with that, top security clearance and all that kind of stuff?

GUY: No, in the space program you don't want to have security clearance. There is nothing in the space organization that is classified. Some people, if they use secret Air Force technology, they have to be cleared. But the space program is in the public domain and there are absolutely no security clearances required in the space program. That way, you can't goof up. They can't prosecute you because you didn't keep all these secrets. Plus, the fact that the stuff that they called "administrative secrets"—they just didn't want anybody to know about the goof-ups they had. They don't want anybody to know how they got the intelligence. They don't want to disclose their sources for fear someone might get hurt.

ROBBIE: How long did they stay at Langley?

GUY: They stayed at Langley until 1962, and then they came to Houston. The big problem was, as I pointed out, where I felt Floyd Thompson's big shortfall was that instead of picking the right people, he got a bunch of people on the basis of seniority, old wind tunnel people, who had no appreciation for space, had no imagination or anything else, but they'd been around for a long time. He promoted them into key positions in the organization at Langley. They just weren't the people that should have been in charge. They didn't understand anything about space. They were aerodynamics oriented, and after all, that was Langley's main mission. Gilruth's outfit, the PARD, was the outfit that was really doing all the pioneering and had all the space smarts.

ROBBIE: Langley's mission all this time was aeronautics?

GUY: Yes. Langley was a real topnotch experimental aeronautical organization. Lewis started out as an aircraft engine research laboratory. It was basically an aircraft propulsion center. Ames was also an aircraft research center. The Ames people were more theoretical, and the Langley people were more the practical people.

ROBBIE: So Langley was aeronautics, and what evolved into Marshall were launch vehicles?

GUY: Yes, but all the other centers tried to get their hands into everything. Ames was aerodynamics, although they tried to get involved in space. They were the ones who launched the monkey into space. Then the monkey died and Gilruth said "God, I'm sure happy they didn't launch him before they launched a man!" *(laughter)* They had this satellite and this poor chimpanzee in there and he was highly upset. And they recovered him, but he died. The program manager was Charlie Wilson, also an LSU grad of our era. This was long after the first man was launched. We launched a few monkeys out of Wallops Island, namely Ham and Sam.

ROBBIE: Did the monkeys wear diapers?

GUY: Yes, they had diapers on, but they were not up that long.

ROBBIE: So, it's 1962 and they all move on down to Houston. You stayed at Langley for another two years and then what happened?

GUY: I decided to leave Langley and go to work in industry. I had an offer that sounded pretty good, chief engineer, stock options and all that good stuff with a medium sized solid rocket company. I was just fed up with the fact that they were trying to reorient us in another direction. We'd been doing great work. They tried to get everybody else in the space business, too. What bothered me was that I had too many activities going and not enough manpower to really make a dent in any of the activities, and yet they kept on putting people in competition with me everywhere else in the organization. I wanted to divest myself of some of these things that I had and put all the people together and let them all work so we would get enough manpower.

I had facilities that would turn out information that no one even looked at. I had so many facilities—like high temperature arc jets. I had guns that would shoot bullets twenty thousand feet a second. I had rockets that would propel payloads into space. I had an experimental rocket plant. I had centrifuges going to check up on some operational problems that occurred only during the flight of solid rockets. I had high vacuum chambers for doing high vacuum research. I had electron beam accelerators to do research on the effects of space radiation on materials.

What else did I have? Oh, I had a thermal optical laboratory to take a look at thermal optical properties of materials that were necessary to find out how much heat stays in the spacecraft and how much gets radiated back in the various materials. I had all these things. But I didn't have nearly enough manpower to get anything worthwhile out of any of them. Then they created the same types of facilities in other organizations who also didn't have enough manpower in competition with me all over the place.

ROBBIE: How counterproductive!

GUY: They wouldn't pull it together. I would have been perfectly willing and happy just to do my thing as good as I was doing it and let them take the manpower and take the facility and get rid of it or at least get something out of it. I couldn't get anybody to listen. They couldn't comprehend.

You see, back in the old days, the way you ran a wind tunnel, a wind tunnel section may have 60 or 70 people in it. The reason for that was to gather data was such a terrible problem. They had to actually use a scale just like Fairbanks Morse Scale. It would measure the force of one of the things with dials. People had to read all that and take notes. They had manometers that would measure pressure from the mercury or colored liquid in the manometers. People would have to read the manometers. They had all these people and then people had to build the models. People had to put the models in the tunnel. You had to have people reduce the data and analyze it.

They couldn't comprehend that we needed more people to analyze the data, and here it is going all to waste. Each facility I had, with its modern high speed instrumentation, could crank out much more data than one of the old wind tunnels. They made richer people poor and yet they were creating organizations to do the same thing I was doing. I decided I had enough of that. I was ready to go off and go to work out in Phoenix, I guess. A rocket company wanted me to come out there and be chief engineer. I was almost ready to take that and then they called me from JSC, where they had had a big shakeup.

Thibodaux Rejoins his Space Task Group Colleagues at JSC

Guy: The project office at JSC had everything in it—all the various disciplinary organizations. They decided, instead of having everybody to work this type of operation, to have the line organizations do all the work and support the project office. They created the Engineering Directorate with all the various discipline organizations which required avionics, and you had instrumentation, propulsion, and power, all sorts of materials, guidance and navigation, flight crew operations, building the space suits and life support systems. They didn't have anybody that they felt could run the propulsion organization. It turned out that I was the second choice at NASA Headquarters, but that didn't bother me. The guy who was first choice didn't come down.

Robbie: They hired you to run the Propulsion Division?

Guy: Gilruth wanted to discuss it with George Mueller. Mueller didn't know me so he had to do some research to find out if I knew something before he would sign off on it.

The Center wasn't built yet. We were spread out in buildings throughout the city of Houston. My office was an old military barracks at Ellington Field across from the Officer's Club. We had a few test facilities set up that were pretty decent. The Center was in the process of being built. We moved in ten months later, I guess. It took us maybe another year to get everything going because it was '65 before we really got going and got everything operational.

We had to do a lot of work! We had a lot of trouble getting the facilities right. The people who built them left them dirty. We had to have them really clean, and not for appearance's sake: the things that we were working with were very hazardous and they reacted to a lot of different things and you couldn't have a lot of junk left around in the pipes. The builders didn't understand the necessities for all that. So, we had to go back and clean and verify everything. We had good people, very dedicated people.

I had all sorts of failures that I learned from. People don't learn from failures anymore because they are so conservative in all the things they do. You used to see rockets plants blowing up and rockets with pieces failing all over. Some of them still blow up. I watched enough rockets spew fire out the sides and blow up that I learned why they blow up and why they don't blow up.

Robbie: And they don't take risks anymore?

Guy: No. The whole thing is, you have to take acceptable risks. You take the risks whether you know it or not. Sometimes you kid yourself about risk-taking. They won't learn from their experiences. For example, we have been flying that shuttle now for 16 years and they haven't learned a thing.

Robbie: What is there to learn that they haven't learned?

GUY: They could learn, for example, that good design matters more than bureaucratic procedure. What Max and I proposed at first was that the solid rocket would be made in one piece, not with these joints that caused the *Challenger* failure. If you built it at a site where you could transport it to the Cape, it would all be in one piece. But that wasn't allowed, because then it couldn't be put up for bids. The only company that could have built it in one piece was right next to the Cape. Anywhere else and you couldn't transport it, because it would be too big. So the whole reason it was cut in half and made in two pieces was so that more than one company could bid on it, because that's the bureaucratic rule.

The most expensive task in a solid rocket is to make the steering right. They had molten aluminum oxide, forty percent by weight of the rocket jet's molten aluminum oxide. You had to protect it from this intense heat, between 5,000 and 6,000 degrees Fahrenheit. You have 1,000 psi pressure, and it has to last withstanding that with no cooling for 120 seconds or more. It had to be able to wobble and not lock up. In order to make it wobble, it's got to move around, back and forth through pressure changes, and there are all sorts of terrible things that could occur that you have to overcome. There is no reason why it couldn't have been designed differently—it didn't even have to be moved. They said, "Well what if one of the liquid rockets goes out?"

Well, if it goes out, then there won't be anything else to worry about. You can forget it because most of the liquid engine modes are generally catastrophic. You shut an engine down to keep it from failing when the instrumentation tells you that its operation is out of tolerance, but you're not sure that it would fail if you kept on going. You could have designed the Shuttle without the need for a movable solid rocket booster nozzle. There are simpler ways to control the thrust vector.

The problem is that the control people overspecify their requirements by a factor of four or five. When we proposed the program, it was to have fixed nozzles. We flew things like that at Wallops Island with no control systems at all. They worked pretty well. That would save billions of dollars in the program because you have to recover the solid rocket.

The back end of that solid rocket has probably got about ten or fifteen million dollars worth of stuff in it every time. I'd get rid of all that.

Robbie: You made this proposal numerous times and nobody listened?

Guy: Oh, yes, no one ever listened. In fact, one of the bidders there bid it that way. They had a good proposal actually, and that would save a lot of money. There are many other things that they could have done that would have saved the program a lot of money.

Robbie: Back in the early days, if you had made that kind of proposal, people would have listened.

Guy: Oh, yes! Back in the early days, I think that I'd have gotten my way. I got my way up to a certain point. Then after that, Gilruth retired. It's funny when you lose your power base. Mine was as a Division Chief. There were many Division Chiefs, thirty or forty division chiefs, but I was one of the IN guys. I was a member of the club. If you don't *know* you are a member of the club then you ain't. Well, you see, it's like the good old boys, sort of like that. You have influence by picking up the phone. You don't write memorandums or things like that. If you want to do things, you pick up the phone and get things done. For example, whenever I'd have problems, Paul was Gilruth's special assistant and I would say, "Paul, how about handling it?" "Sure, I'll take care of it."

Robbie: You told us that Gilruth used to go around to talk to all his employees and sit on the trash can without turning it over at each one's desk to check on how he was doing and find out what he needed.

Guy: Yes, once a week he would stop by. I used that technique over at my Division at JSC all the time. I knew everybody in my Division, and everybody knew me. That's the difference. I got along fine. I could tell you something about everybody and their families and everything in my

Division. I found out I could go and talk to them in their offices much better—they felt much more comfortable than if I called them up and asked them to come and talk to me. They would all wonder, "Well, what the hell is he going to want?" So I would go to them, and sit around and listen to them. I'll tell you what I was doing—I was leaving post-hypnotic suggestions, is what I did when I went around.

ROBBIE: What do you mean, "post-hypnotic suggestions"?

GUY: Well, I would do to them what Gilruth did to me. When I wanted them to do something, I didn't have to tell them I wanted them to do it. I would leave a post-hypnotic suggestion. Then they would go ahead and do it that way, maybe even thinking it was their own idea.

ROBBIE: What year did Gilruth retire?

GUY: Right after the Apollo program, I guess. Before the Shuttle started. He decided he didn't want to get involved.

KEN: In those early days when everything was going great, did you have any premonition that in the future it would become more bureaucratic?

GUY: Oh, yes. Hugh Dryden was very astute—he was very concerned about the NACA becoming the NASA. He thought the whole flavor of the organization was going to change. We didn't start going downhill until about the seventh year. I began to sense something about the seventh year, that the agency was falling.

KEN: Now, let's see. The space agency was formed in 1958, so we're talking mid-60s here?

GUY: Yes, Apollo was on its way, but I could sense that things were changing. We were getting a different class of people in charge all of a sudden. I sensed a tremendous change when Mr. Webb left. I've been retired for 17 years and what have they accomplished in the 17 years I've been retired? We've flown into a Russian space station and we launch almost all of our commercial satellites on foreign launch vehicles now. And we're buying rocket engines from foreigners. That's what we're doing now. No one can dispute that.

Robbie: In other words, you don't regret retiring when you did, because JSC hasn't done anything significant since then anyway.

Guy: No, nothing significant. They became an experiment in social and political change and lost sight of what they were supposed to do. Their bureaucracy took over the White House staff that basically controls all personnel who run NASA. The Congressional staff controls all the programs that NASA does, and the lobbyists, and everything else. It just got to be where it wasn't any fun. Most of the other people I know would say the same thing.

Robbie: How long were things good at Johnson Space Center? How long was it before the bureaucracy took over and you felt that you were not happy?

Guy: It was the beginning of Apollo. The shuttle program was passable, but it lacked a lot of real talent. The people in charge felt they were working for industry rather than for the agency. Their decisions were based pretty much on what industry wanted—a lot more than what *we* told them to do. There are a lot of ways we could have saved us some men and a lot of money. The Air Force got involved with requirements and other big operations, which they called "flight operations." They got to be so ultraconservative that they put all sorts of ridiculous requirements on things. Then they won't watch what's going on and see that there are things you can change here. It changes your attitude, really. That is how you go about things. Contingencies, for example, and redundancy, and many of those things. You would have to use it even if you could prove you didn't need it and that you could simplify the system.

CHAPTER 2

GUY THIBODAUX
MAX FAGET · PAUL PURSER
THE SPACE CADETS

BIOGRAPHIC INFORMATION FOR MAX AND PAUL

Maxime Faget (August 26, 1921 – October 9, 2004)

Max Faget graduated from Louisiana State University with a BS degree in Mechanical Engineering (aeronautics option) in June 1943. He immediately joined the Navy as a reserve officers assigned to submarine service. He then became employed at the Langley Laboratory of the National Advisory Committee for Aeronautics (NACA) in August 1946. He was assigned to the newly created Pilotless Aircraft Research Division (PARD), a division that was to fly rocket-powered models of aircraft and missiles at transonic and higher velocities to obtain aerodynamic data. During this period he did pioneering work on supersonic inlets and ramjets. He designed a compact (6½" dia.) ramjet engine and a supersonic flight test vehicle, which was powered by two of these ramjets. During a flight test in 1950, this vehicle accelerated under ramjet power in a climbing flight, achieving an altitude of 65,000 feet and a velocity of M=3.2, setting unofficial speed and altitude records for vehicles powered by air-breathing engines.

While at Langley, Max was appointed to the four-man team that prepared the conceptual design and performance analysis of a research aircraft that could fly twice as fast and much higher than currently possible. The NACA then approached the Air Force to contract with industry to complete the design process and manufacture such an aircraft. This became the start of the X-15 program. Faget was also appointed the NACA member of the Polaris Missile Steering Task Group, where he proposed the aerodynamic shape that was employed for the reentry warhead. During the winter of 1957–58, he conceived the design and started development of the one-man spacecraft subsequently used in Project Mercury. Both the Gemini and Apollo spacecraft are derivations of the Mercury concept. When NACA was notified that it was chosen as the cadre from which a new civilian space agency would be formed, Faget was appointed a member of the transition team.

After NASA was formed, Faget was assigned to the Space Task Group (STG) organized to manage Project Mercury. Although Mercury was the main task at STG, there was great interest in follow-on programs. Consequently, he devoted a large part of his time to heading a design and analysis team exploring manned flight to the vicinity and the surface of the Moon. Because of this and other NASA studies, President Kennedy was able to commit the USA to a lunar landing by the end of the decade. With the advent of Apollo, STG became the Manned Spacecraft Center (MSC). Faget was appointed Chief Engineer at MSC, responsible for the design, development and proof-of-performance of manned spacecraft and their systems. This responsibility included specifying the function and design of numerous engineering laboratories to be constructed as part of the MSC. In April 1969, shortly before the first lunar landing, Max organized a special preliminary design team to do an intensive feasibility study of a reusable manned spacecraft. This effort achieved program status when Johnson Space Center (nee MSC) was given formal authority to develop the Space Shuttle. Subsequently, Faget gave prime emphasis in his personal activities and those of the organization toward solving the manifold problems in the development of the Shuttle. He retired from NASA after the Shuttle successfully completed its second test flight in 1981.

In 1982, Faget and several Houston businessmen founded Space Industries Inc. (SII). SII designed the Industrial Space Facility (ISF). The ISF was to be a manned-tended orbiting facility to be used for experiments in a high-quality micro-gravity environment with special emphasis on material processing. Westinghouse became a partner with SII in a joint venture for financing, construction and operation. Significant backing was obtained for an initial deployment in 1992. However, the aerospace industry, Congress and NASA feared it would curtail the Space Station program, and consequently the ISF was never deployed. SII then manufactured a wide range of experiment support equipment that was flown on numerous Shuttle missions. The most significant was the Wakeshield built for the University of Houston. This free-flyer was successfully deployed on two missions, providing the experimenters with an ultra-high vacuum environment for material processing.

Paul E. Purser (Dec. 9, 1918-Sept. 8, 2003)

PAUL PURSER was born and raised in Southeast Louisiana, and he graduated from Louisiana State University in 1939 with a BS in Aeronautical Engineering. He took the Civil Service Exam that spring in anticipation of a major growth in NACA (National Advisory Committee for Aeronautics) during the pre-World War II period, and worked briefly as a Junior Inspector at Glenn L. Martin, Co. Baltimore on a twin-engine attack bomber being built for the French armed forces. In mid-October he received his NACA appointment and immediately departed for Langley Field.

During the prewar and WWII period, Purser carried out wind-tunnel research and evaluation work on practically every aircraft proposed for, or used in, World War II. At the end of WWII, he joined the Pilotless Aircraft Research Division (PARD) which was then being formed at Langley. There, he and his colleagues carried out aerodynamic and structural research in flight using rocket-propelled models at Mach Numbers up to 15. Additionally, Purser headed up the development of various high-temperature ground facilities for research on materials, structures, and missile nose cones. During the major portion of the period between October 1957 and October 1958, he was a member of the small (12-man) team that conceived and "sold" the U.S. Manned Space Program and as part of the larger (75-man) team that planned and implemented the conversion of NACA to NASA.

From October 1958 to April 1970, Purser was Special Assistant to the Director of Project Mercury, which developed into the Manned Spacecraft Center (now Johnson Space Center). During that same period, he continued his membership in the American Institute of Aeronautics and Astronautics (AIAA). In the early 1960s, he became a charter member of the American Society for Oceanography and a Member of the Marine Technology Society when the MTS and ASO merged. Also during the 1960s, he was registered as a Professional Engineer in Louisiana and Texas. Purser was also invited to (and did) join 3 National Honor Societies: Tau Beta Pi (Engineering), ODK (Leadership), and Sigma Gamma Tau (Aerospace Engineering.). During the 1968–69, academic year he was on loan to the University of Houston where, as Special Assistant to the President, he guided the development of the UH-Clear Lake Graduate Center.

Upon his retirement from NASA in April 1970, he began consulting in various fields, trying to apply what he had learned in NACA and NASA. In addition to various consulting tasks in the oil & gas industry, this led to a 10-year stint as Staff Consultant to the NAE/NRC Marine Board, overlaid with a 5-year stint as consultant to the Stanford School of Medicine Cardiology Division, about 5 years as a part-time Systems Engineer with the Gulf Universities Research Consortium (GUBC), and a 25+-year association with Capt. W.F. Searle (USN-Ret.) on various tasks in the oceans industry. During this time, he prepared and presented several technical papers at the Offshore Technology Conferences and the MTS Oceans Conferences.

INTERVIEW

Recap: *Interview #1 took place with Guy Thibodaux, the engineer and propulsion scientist responsible for the propulsion work on Mercury and many other space projects, at his home near Johnson Space Center in Houston/Clear Lake, on Sept. 9, 1996. The following day, Ken and Robbie met again at Guy's home, this time to conduct a joint interview with Thibodaux and his colleagues Maxime Faget, who was instrumental in the design of the Mercury, Apollo, and Gemini spacecraft and the early Shuttle, and is widely considered the father of spacecraft design; and Paul Purser, engineer and manager at Langley Research Center in the Pilotless Aircraft Research Division (PARD), which formed the early nucleus of the space program.*

At the Beginning: Louisiana State University

Robbie: I know that all three of you went to LSU, which is remarkable—three key figures in the development of our national space program came out of the same university, even the same engineering department! Obviously, something went on at LSU that produced the three of you—what was it? Paul, why were you there?

Paul: Because that was the college that I could afford. This was in the middle of the Great Depression. My family would have liked for me to have been a lawyer, or a doctor—or even a dentist would have been acceptable. But no, I wanted to go with what the Dean of Engineering at LSU referred to when I got over there as "aeronautics engineering," a synonym for "artistic hunger strike!"

But I was hard-headed, and I was going to be putting up most of the money from state funds and various other places that I could get jobs, plus a double work scholarship where I would work 100 hours a month in the engineering labs for 25 cents an hour. That was enough to take care of me. And my tuition, $62.50, was the right amount per semester. On one year of junior college, I earned everything by driving the family car 15 miles from home to school with four or five other students who paid me to drive them. I had enough out of that to buy gas and oil for the car and buy my lunches and cigarettes and pay the little bit of tuition that I had to pay the junior college. When I got to LSU, I had a little bit of money from home. My job was to work in the Department of Aeronautics lab and that was the place that I really gained an appreciation for what technicians mean to the success of engineers.

ROBBIE: So you specialized in aeronautical engineering at LSU. And what year did you graduate?

PAUL: I graduated in 1939.

ROBBIE: And did you know Max and Guy while you were there?

MAX: We graduated in 1943.

After LSU: Langley, Here We Come!

ROBBIE: Paul, when you left LSU, where did you go?

PAUL: Well, Walt Williams and I took off and came up East, hunting some kind of job. We had both taken the Civil Service Exam hoping to get on at NACA. That was because in the late 1930s, Charles Lindbergh went over to Europe to see what was going on in aeronautics there. He came back and told everybody what an air force the Germans had built. And the press almost got him run out of the country, because they thought he must be a Nazi to come back praising the Germans like that. He wasn't praising the Germans, he was answering the question what were *they* doing compared with what *we* were doing. This was a time when Congress and the powers that be in Washington decided to listen to advice from people who knew what they were talking about. At any rate, Congress decided that the U.S. better get busy, and that NACA, which at that time had about 700 employees, should expand markedly and steps should be taken to increase education in aeronautic engineering all over the country.

So Henry Reid, the engineer in charge, wrote letters to all of the academic people he knew, particularly those who had worked with him, and said, "Get your best seniors to take the Civil Service exam because we are going to be growing like mad. "

Around October 1, I got called to NACA and that same day I went home and packed my clothes and caught a bus and headed for Langley. About six months later, Walter got his offer and he came down.

ROBBIE: What was your initial position at Langley?

PAUL: I started out as a junior engineer in the atmospheric wind tunnel. It was a seven-foot high, ten-foot-wide structure that tested both complete models of aircraft and wing sections with various controls.

At the end of World War II, John Stack wanted more space to build supersonic wind tunnels so we could try to catch up with what we found out the Germans had been up to during the war. So they closed down the wind tunnel that I was running and gave the building to Stack to put in some small supersonic tunnels. I was then offered the opportunity to either go into another subsonic wind tunnel and run it, or go to work for Gilruth with the screwballs he had, fixing to fly rocket models off Wallops Island. And I decided well, I know enough about subsonic wind tunnels, I'd like to learn about this other stuff. So, I took the job with Gilruth—was one of the guys in PARD.

GUY: You didn't tell them about your stint in the administration office.

PAUL: *(laughs)* During the very early days just before WWII, they couldn't decide whether they should give the draft deferments to

NACA or not. So, they had the War Manpower Commission conduct a study of what it would take, person by person, to retrain and replace every male employee throughout the whole agency. There were about six or seven hundred of us at Langley at that time. And somehow I got chosen as the host of the War Manpower people who were going to conduct the study, which meant that I was the local guy who was going to do the work and they would sign the report. *(Laughing)*

At the end of the War Manpower Administration study, I had a really good understanding of what was going on at the NACA, and someone higher up decided I knew enough about the whole picture to be the Laboratory's Budget Officer. I accepted the job for a six-month trial period to see if I liked it. At the end of six months, I knew I preferred research and I was allowed to go back with no loss.

The study itself concluded that it would take a total of eight to ten years to replace everybody by training other people up to the level of experience that the Langley people held. So, they decided they were not going to draft *all* of us. But they got hold of the services, and the Air Corps agreed to take the people at Langley and at Lewis Lab, and the Navy took the people at Moffett Field, because Moffett Field was a naval installation. So, as our numbers came up near draft status, we would go ahead and get drafted and be inducted.

We'd go up and take physical exams and line up, and they'd say, "So-and-so and so-and-so, you are hereby inducted into the Armed Forces of the United States of America and assigned to the Air Corps Enlisted Reserve on inactive status for the duration of the war or until the time that you leave your job at NACA."

ROBBIE: Good incentive to stay!

PAUL: So, I became a veteran with no active duty time *(laughs)*, because the ten seconds that it took to read that sentence were too short to list, as active duty time was listed in years and months! At the end of the war, I got discharged, got a sheet of paper that said, "You are hereby discharged with honorable status."

Guy and Max were getting discharged at about the same time, but they had actually done some fighting. They went by LSU to say hello to Fritz and other friends, and to find out what Fritz was working in on. He said he thought that if they wanted some interesting work, that they ought to come up to Langley and talk with me to see if there was a place there for them.

So these two boys came barreling up with their pockets full of discharge money and accumulated leave money. PARD had reached a stage where they wanted a good chemical engineer guy interested in solid rockets. And they wanted a good mechanical engineer who was interested in ram jet propulsion and liquid rockets. So Max and Tibby came in, and they got officially greeted and put on the list as "visitors seeking employment." We sat and chatted for two, three, four hours. At the end of the time, I said, "Well, I don't think we have any problem at all hiring you guys."

GUY: We never looked for another job, by the way, Paul. *(laughter)*

MAX: Absolutely! Never even *thought* of getting another job! I don't know who brought it up first, but we started going up north, and I guess we'd been in the car maybe half-an-hour or an hour, and that's when we decided "Well, what the hell—we can have a lot of fun now!" *(laughs)*

"The NACA Nuts": Right People, Right Place, Right Time

ROBBIE: Who called you the NACA [Guy: pronounced "nakker" as in "cracker"] nuts?

PAUL: All the people around the town of Hampton and in the population centers around Langley field. And some of those NACA folks were *really* nuts. There was one guy who had a piece of white adhesive tape on the top center of his steering wheel and he would drive along solving mathematical equations in his mind and just glance down once in a while to see if the white was still in the middle! *(laughter)*

There was a hardware store in town that guaranteed their products. They had ice cream freezers and some guy bought one that was guaranteed not to rust. About six weeks later he brought it back all rusty, so the hardware store owner replaced it. This was repeated about six weeks later. The third time the owner was asking, "What are you doing to these things?" "Well, I was just testing the guarantee out. I filled it with salt water and put it out in the back yard." So he did it every time for about six weeks. The hardware store owner said, "I can't afford this guarantee for you. I'll just give you your money back." *(laughter)*

GUY: Paul Hill, our first boss, was a really nice guy who liked us and he did everything for us. Paul would go see a movie, and if he liked the movie he'd buy tickets for us and he'd give them to us. If he had a book that he liked, he'd buy us a copy of that book. He had a sailboat. He'd take us out on the sailboat and go sailing all the time. He was really interested in us as people and in helping us get along. He was kind of a strange character, very, very strange.

MAX: He wrote a book on Unidentified Flying Objects.

PAUL: I didn't know that he wrote a book, but he was one of the first ones to report on seeing any in the Langley area.

GUY: You can't find three people who were happier or more interested in the work we did. We happened to grow up at a very challenging time in world history and happened to work for the right organization.

We had the right people, in the right place, at the right time. They were *real* people. No facade, no ego. If they had any egos, they never let them come out too much. We all worked for the good of the organization. If you did a good job at the organization, it took care of you. That's the way it worked. No awards. The reason I retired was that it quit being fun anymore.

The X-15

MAX: The NACA sold the Air Force on a lot of good projects. They'd say, "We don't have the money to build an airplane, but we would certainly like to do this kind of flight research. Let's do a joint program." The Air Force was always very interested, because we had some pretty nice toys, you know? *(laughs)* We'd give them the idea for another good toy, and they'd say, "Oh, yes. We'd like to play with that one." That's how we did the X-15."

The case of the X-15 is very interesting. The Air Force contracted with North American for the X-15. What did NACA do? They made paper studies. We had a team of four people, who were put together to do these studies. I happened to be one of the four, and I was in charge of performance, and I needed help on propulsion, so I'd go back and Guy would tell me what kind of engine we ought to use.

GUY: We called the shots on the rocket engine and what propellant it would have—much to everybody's consternation, we made the decision on that.

MAX: We had another guy who was doing aerodynamic design and he came up with the wing loading and where the wing was going to be and things like that. The funny thing about that is, I couldn't figure out how much it was going to weigh, so I got a hold of some stuff—like wings weigh so much per square foot, and didn't weigh so much for skin areas. And it turned out that the X-15 came within about 10 percent of what I predicted it would weigh. I've never done that well since! *(laughs)*

But we didn't know how fast it was going to go, so I ran a bunch of different trajectories, and it turned out that there was a kind of hump in the curve. You'd try to go much faster than Mach 6 and the vehicle was starting to get pretty big, so I recommended that we do Mach 6, and everybody said, "Oh, yeah, that's a good enough number, Mach 6." *(laughter)*

Sixty thousand feet a second was actually what I recommended, not Mach 6, but it comes out about the same. And I think it actually did do just about that. Then NACA went to the Air Force and said, "Look here...you can build an airplane that is going to be this size and have this weight, and you can carry it on the B-52." We said, "Oh, gee, that's gonna be a lot of fun, we really like that." Then the next thing we negotiated was, "Well, who's gonna fly?" They ended up just saying, "Well, there are going to be NACA pilots who are going to fly it and Air Force pilots to fly it and Navy pilots to fly it."

Guy: North American pilots? Scott Crossfield?

Max: They wouldn't let Scott Crossfield fly it to anywhere near its maximum performance envelope. He only flew it at low speeds to prove that it'd land. He was only allowed to fly to prove that he could land and operate the engine. That was an acceptance flight. Then performance flights were done by the Air Force and other government pilots.

And the next toy we were going to design for the Air Force was what Round Three was supposed to be about. Instead of going Mach 6, it was going to go at least Mach 12 and hopefully Mach 15.

Paul: It was going to be called the Dynasoar.

Max: So that was going to be the next project and that's what we went to Round Three to discuss.

The Round Three Conference

Max: At Round Three, the people from over at Langley, the people from Ames, the people from Cleveland, as well as from the Flight Research Center, were all going to be there discussing what this beast ought to look like. And we all had different ideas.

Guy: We were the practical people. The other people were the eggheads.

Max: They were real eggheads, they really were! *(laughs)*

GUY: I think that the Round Three Conference is where we got started in the space program. Do you all agree with that?

MAX: It made a lot of things legitimate. The thing that bothered me about the Round Three Conference was that I thought we were going to build something that was going to go fast, an airplane-like affair. And I came back convinced that we weren't going to do that.

KEN: Now, at this point there wasn't a civil space program. These were all military considerations at the start.

MAX: Right, and there wasn't anyone really thinking about going into orbit except possibly for reconnaissance—that was one of the long-range things the Air Force was thinking of. The Russians put Sputnik up before Round Three. It scared the hell out of everybody because we were working on something that was going to put a grapefruit up and they had already put up a gallon jug!

PAUL: It was actually *during* the Round Three Conference. I don't think it was even a week before. I believe it was either the day before or during, because the news got to us when they were sitting there in the auditorium at Ames.

MAX: It was kind of a shock, and it changed the rules. As far as we were concerned, up until Round Three, it was a toy — because we were looking at trying to get new technology, the toy was going to be this ROBO and it was going to be a lot of fun to get involved with all this. The military said," We don't want that." So here are the Russians putting up another toy for us. So during Round Three, we got to thinking about getting up into orbit as opposed to discussing glide bombers.

KEN: So *that's* how the move into orbit first got started.

MAX: Yes, *within* NACA, with just a very few high-level people. Prior to that, we heard about the Naval Research Lab, which had the Vanguard program to launch satellites.

PAUL: It was for the International Geophysical Year Scientific Payload.

MAX: Mr. Abbott from our Washington Office came to Langley for a visit, a month or so before Round Three, and we were all sitting around talking. We had just put up a rocket that had gone Mach 15. In fact, during the last year, Bob Piland had put up about two or three Mach 15s. And we got valuable data from Mach 15. Meanwhile, there was a new class of larger rockets coming in. I got to talking to Guy, and Paul and somebody brought it up. I said, "With five-stage rockets, I think we can put something into orbit probably before Vanguard does."

GUY TO MAX: Dryden had told him that the Navy was the only one who was going to be allowed to put a satellite in orbit, and that *we* couldn't do it. And then Max sprang the manned thing on him.

MAX: It wasn't that great of an idea. You just look at the rockets that are available. You can pick the number of rockets that you think are going to be available and you stage properly. Each rocket would be maybe one-third the weight of the previous rocket, or somewhere between half and a third the weight of the previous rocket, and if you stage enough of them together, you can go up to Mach Infinity if you have enough stages. In our assembly, we had made five-stage rockets. All of a sudden, a string of rockets showed up that indicated to us that we could get up to orbital velocity.

KEN: But at this point you hadn't jumped to the manned space craft.

MAX: Oh no, no.

KEN: This is just an early part of "Put something up there." National prestige versus the Russians and all that.

Eggers, Allen, and the Manned Spacecraft

MAX: Eggers got us into the manned spacecraft. With his inside knowledge, Eggers had been working on putting a man in orbit. He was not worried about what was going to put this guy in orbit, he was putting his attention to what kind of vehicle this man was going to fly in.

KEN: The spacecraft and the reentry.

MAX: He'd apparently put in, I guess, at least a half a year's work on it, because he had a fairly well-thought-out idea for a lifting-body spacecraft.

KEN: Eggers was NACA?

MAX: Yes. He was out at Ames. He had come across what he thought was a neat idea, and it *was* a neat idea. You build something that uses solid-rocket fuel. You take a basic shape, which is a cone with a hemispherical nose on the front of it. Then, you slice it in two and move the center of gravity to the right place, and it's stable. So it's not only stable, but it's a stable body giving lift. It was a great idea looking for a market.

PAUL: In one of the preparation talks for a major nationwide industry, government, academia conference on the current or newest state of the art in a specific area of research which we used to have one or two of every year, some bright young guy from Ames got up and made a very learned statement about something. Harvey got up and essentially said, "Baloney." And went ahead and gave a simple explanation of what it really should be, and the young guy got very upset—"You won't let me have my say." Harvey said, "I did let you have your say and then I had mine. I am a firm believer that everybody in this room has a right to have his say. And I will defend that idea until your death!" *(laughter)*

MAX: Harvey was a person in the NACA who has received quite a bit of renown. He had two hobbies. One was that he was a Maya fan. Every time he would get a vacation he would go down there and look at all these old pyramids. His other hobby was gathering meteorites. And he noticed that the little meteorites could get all the way to the ground, but they were pretty small, and that they had a rounded nose. It had become round because the surface had melted, and it had melted in the position that made them, apparently, aerodynamically stable, as far as he could tell. So the meteorites had an aerodynamically stable hemispherical nose on them with a small after-body, which wouldn't have melted. And he got to looking at these things and he said, "The reason they survive is that they have a very high drag-to-weight ratio." So he decided that the way to get down from orbit is to slow down!

KEN: As you go through the atmosphere, you limit the heating by slowing down.

MAX: So I was at Ames listening to both Eggers and Harvey Allen. I've always liked simple things, and I kind of liked what Harvey was saying. So when I went back to Langley, I first wanted to check on how much heating the lifting body had versus the non-lifting body. I put two guys in my branch working on this issue to compare these shapes. We wanted to see which would be better. Of course, the full body would have a pure ballistic entry, whereas the half body would skim along the top of the atmosphere.

We found out two things. One of them was that even a pure ballistic vehicle tends to skip a little. It doesn't actually gain altitude. It tends to skip a little. It doesn't ever rise but it stops falling so fast. The lifting body definitely skipped, but if you used the right angle of attack, you could keep it up in the air. Of course, Eggers recommended this to control its cross range. I think that was his idea.

GUY: None of these things had any real control. They were all passive reentry bodies.

MAX: Except that Eggers wanted to control the cross range. And sure enough, there was twice as much heat on the thing with lift as opposed to the one without lift, because it was in the atmosphere longer. A non-lifting body would end up maybe hitting 8 Gs and the lifting body only a maximum of 2 or 3 Gs. If you are decelerating at 2 or 3 Gs, it's going to take you longer to decelerate than at 7or 8 Gs. Intuition tells you it's going to be better to have the drag. And Harvey said it was better. Harvey wasn't particularly keen on Eggers's idea because he thought, *w*hy don't we just slam into the atmosphere? You are trying to get out of orbit, after you've been up there long enough, enough is enough, get down. *(laughter)* You can't argue with that logic!

KEN: Provided you don't burn up.

MAX: At Langley, we did an analysis, and one of the guys, Ben Garland, said, "If you turn it around and enter the thing backwards, with the cone in the rear instead of the front, you are going to have a lot more drag." And I said, "You sure as hell will!"

GUY: The last time I talked to Al, he says he still thinks you came in with the wrong end first!

MAX: I know that! *(laughter)* Anyway, the complete body of revolution that Eggers starts with before they sliced it in two is called a Discoverer shape capsule. And actually, that's the shape of the early warheads that the Air Force would use.

But as it turned out after we got going, the Air Force proposed a manned ballistic vehicle with the Discoverer shape. You know what? They had a gimbaled pressurized sphere, mounted within the Discoverer capsule because they came in the wrong way. They had to position the sphere so that the astronaut would be taking the drag eyeballs-in going outbound, then they'd flip him over 180 degrees for the entry drag. He would then enter eyeballs-in. I just flipped the whole thing, which was a hell of a lot simpler and about *half* the weight!

The Mercury Capsule

MAX: You can understand how a spherical segment front end would be stable, because the pressures are focused at the center of the spherical surface. Since this point is well behind the center of mass, we get something really stable. What it is not, however, is dynamically stable. It has the dynamic stability of a falling leaf. I've got to illustrate this because I love to do it. [Max went outside, picked up a leaf, held it high, and then dropped it. Of course, it didn't fall in a straight line, but rather wobbled and wafted.] Woody Blanchard got very interested in all this and said, "You know, they have a spin tunnel over here, which is nothing but a vertical tunnel. Let's make models that have enough drag and are sufficiently light so that they will float in the test section. Then we can change the length of the conical after-body until we get something that may wobble back and forth but will not topple."

KEN: What you are describing is what you did that was new that the Air Force didn't do when they went through their ballistic missile. When their target came down, they didn't consider any of this. This was new territory now.

MAX: You see, what happened was there was a big rush [because of Sputnik] and people weren't thinking the problem through, particularly at the management level. The Air Force had a reentry body that they used as a ballistic warhead. It was both statically and dynamically stable. They understood the heating on that reentry body. They knew that if they scaled it up and if they reduced the angle of entry that the heat load per square foot would go down. So they thought, "Well, hell, that's good. That solves the entry problem. We'll just use that. Don't bother me with anything else. We've got a thing that works!" So, that was what they decided to use.

"Now," they said, "we're going to put a guy in there." They ended up with something that was estimated to weigh about 4500 pounds. And our original estimate on Mercury was 2000 pounds.

We put this Mercury capsule in a spin tunnel and made a scale model about 8 inches around, then we added various conical after-bodies making them longer and longer to reduce the dynamic instability as more and more surface became exposed to the oncoming air. Finally, the oscillation was decreased to about plus or minus 60 degrees without tumbling. And we said, "That's good enough!" *(Laughter)*

That's how the length of the cone on Mercury was established. We knew that it would always have to get to some angle of attack before the cone would get enough aerodynamic force on it to limit the oscillation. Now the vehicle was designed so that in the event that the attitude control system failed—and we *expected* that attitude control system to fail—we would still make reentry.

Who Ran the NACA? A Bottom-Up Organization

MAX: NACA would always study what the customer wanted. And the customer was the Air Force, the Navy, and a little bit of commercial aviation. But, we always, in our pride, said, 'Yes, we'll look at what you want, but we will always do some basic research to investigate the problem further." That's what we were doing.

ROBBIE: Did you have a mandate to do that or were you just bootlegging?

MAX: Oh, we had a mandate. Our mandate was that anything the guys down in the lowest part of the organization thought was a good idea to follow probably had merit. You'd talk it over with your supervisor—if he thought it was a good idea, you'd get it funded. *It was a bottom-up organization like you never saw before!* It was a bottom-up organization with a big free ticket. It was a wonderful place to work.

PAUL: The ticket wasn't very big! We still had to go to Headquarters to spend anything over $5000.

MAX: Yeah, but that wasn't my problem! *(laughter)*

PAUL: I got my experience in the R & D bureaucracy during the war. Guy and Max got theirs on a jeep in Burma and on a submarine in the Pacific. We were all equally mature but more experienced in different things. We all knew: who runs the Navy? The chief petty officers! And who runs the Army? The sergeants! And who ran NACA? *Us*! Dr. Dryden was smart enough to realize that.

And he was personally not in favor of NACA becoming NASA. But, he did listen to the input from the people in the Centers who knew what was going on. So he was the primary salesman for having NACA chosen as the new place for NASA, even though he really didn't believe in it. But that was just his personal opinion. His professional opinion, based on the advice of the people who knew the business, was positive.

And that is how NACA grew up as it did, and then had to turn around later when the fount of all wisdom became centered in Washington, and the Centers were told, "Do what we tell you to, write it down the way we tell you to, and don't bother us with these screwball ideas."

ROBBIE: So that was when it became a top-down instead of a bottom-up organization. At what point did that happen?

GUY: It began to change after Apollo, but through Apollo it was all bottom-up.

PAUL: Well, they *thought* it was top-down!

GUY: Well, yeah, they *thought* it was top-down. They never understood what went on!

PAUL: And they created lots of trouble by thinking that. But in spite of what they thought, it still actually worked from the bottom up.

MAX: And meanwhile, up at Headquarters, they were building an organization to tell us how to do it. Joe Shea was up there in that organization. He'd come down and talk to me and I would say, "Yeah, ok, but we're going to do it this way," and he'd get real frustrated. *(laughs)*

So Gilruth was looking around for somebody to replace Charlie Frick, who had resigned, and Joe came down and volunteered. I said, "Joe, how come you came down here?" He said, "You convinced me! Apollo is being run down here in the Center and I want to be part of it." *(laughter)* He said," I am tired of fighting you guys and getting nowhere!"

PAUL: He hadn't learned his lesson completely, because he was still the top dog when he got here. We still had to battle him just to start—just a few minor things, like having a little bit of flammable material in the oxygen atmosphere of the Apollo Command Module. Joe insisted on approving a little bit of flammable material in the Apollo Spacecraft several times in spite of the continued advice to the contrary of Faget's materials specialists and others.

MAX: He had to know everything that went on, down to every minute level of detail from everybody. I think he missed the big picture because he was looking so much at the detail.

PAUL: But the problem was he not only missed the big picture, he didn't comprehend all the details of each organization. He was hearing but not listening.

MAX: He never changed.

The Early Space Program: PARD as the Center of Gravity

MAX: I would say in the first three months of 1958, the country acted in a very panicky way in reaction to Sputnik. Still, the first three months in 1958, we didn't dare think so big as saying, "We're going to be *the* agency." Like everything else that we came up with, we tried to sell it to the Air Force. People got very interested in the orbiting spacecraft, and we put together a dog and pony show and went to see various people in the Air Force, trying to sell them on this as a good project. We didn't know then that this was going to be a civilian space program, and that *we* were the ones who were going to make it happen.

KEN: So what you've really said in this critical three-month period is that the three centers were Langley, Ames, and Lewis, and Ames didn't choose to get heavily involved, so it was some combination of Langley and Lewis that formed the basis.

MAX: And Lewis, of course, was the propulsion outfit, except they were not in rocket propulsion. *(laughter)* We knew so much more about rocket propulsion than they did, it made them sick!

KEN: So the real center of gravity was right at Langley.

MAX: It was right in PARD! Let's not kid ourselves. It was right in that one little division. No doubt about it! Everybody likes to gloss over this, but that is a fact. This one division, perhaps a hundred people within the NACA, was the focus, the eye of the storm, if you wish. Everything rotated around that. We were just lucky to be there—and clever enough to take advantage of it! *(laughter)*

PAUL: Gilruth and Tommy Thompson. Both had an ability to get anywhere from two to twenty people together and let each one of him have his say. And at the end they would say, "You know, we can conclude that we ought to do such-and-such." And nobody in the room could disagree. Not because they were knowledgeable, but because they got the essence of what ought to be done from listening to everybody, putting it together, and at the end, summarizing it very briefly. And everybody would go away happy.

The Space Cadets, the Air Force, and the Pentagon Meetings

MAX: Around March of 1958, the President's science advisor, James Killian, talked it over with the President, and the President said he wanted to have a *civilian* space program. Eisenhower was very concerned about the military-industrial complex in this country. He made a number of speeches about it. He was concerned because in Germany, the military-industrial group took the country over. And he knew this. It led to dictatorship. He was certainly very strong on this point. Killian told him that "the NACA is your best cadre." So they got Dryden in there and they talked with him. And all of a sudden there was a big panic at headquarters, because we had a bottom-up organization. And Headquarters was *absolutely unprepared* for the idea of running something like a space program—absolutely! *(laughter)* So, Dryden and Crowley got hold of the Centers and they said, "Please send some people up here to help us get ready for the transition."

PAUL: The Space Cadets were on the sixth floor of the Dolly Madison building.

ROBBIE: What Space Cadets?

MAX: Us! *(laughter)*

PAUL: That's what they called us—the Space Cadets! *(laughter)*

MAX: There are the three of us, and probably not more than two or three other survivors. We were it!

ROBBIE: Did you call yourselves the Space Cadets at the time?

PAUL: The *people* at headquarters called us the Space Cadets. Now, what the *officials* at Headquarters called us, I don't know. *(laughter)*

GUY: There were just a few guys who knew what we were there for. We had a whole floor and they weren't allowed to talk to us! *(laughs)* Everybody knew that we were going to possibly transfer the space program. It wasn't that secret. They didn't know what was going on, what we were doing in detail.

PAUL: It wasn't a question of confidentiality. It was just that we didn't have time to explain it to people. We were too busy doing it!

MAX: It was a completely ad hoc organization pulled together essentially by saying, "We think you could help!" "OK, I'll volunteer." You could say, "No, I don't want to do that." Ames did not send anybody up there because the travel time was too long between Ames and Headquarters.

MAX: Lewis was about the same distance from headquarters as we were. Maybe just a little further, but hardly noticeable. We'd go up there on a Monday or Tuesday and come back on Friday.

GUY: We used to run through the Pentagon like we owned it—I don't know how we found out about those meetings. Zimmerman or someone gave us a list where all the meetings were at the Pentagon. And somehow the ground had been cleared for us to go, and we'd go to any damn meeting that we wanted to over there, and either be quiet or participate.

MAX: I spent most of my time over in the Pentagon talking with the Manned Space Panel. The Department of Defense—I think there were five or six people on the panel. NACA had two, including myself.

MAX: Of course, in those days, the instrumentation was not as good as we have now. He had this one chart that showed—one of the instruments they were going to have in the control room was called a Death Meter.

KEN: A Death Meter?

MAX: They combined about three or four parameters, and some of these parameters in the best medical understanding of the Department of Defense showed how close the guy was to death: his breathing rate, his pulse rate, deep body temperature.

PAUL: When your temperature gets low enough, and your respiration rate gets low enough, and your heart rate gets slow enough, you're going to die.

MAX: They were going to watch the Death Meter in case the astronaut got too close to death *(laughs)*, and then we were supposed to bring him down, I *think*!

PAUL: Nobody knew, really, *how* these things all were going to run.

MAX: No one was going to run experiments either! *(laughter)*

PAUL: And any combination could kill you at a higher level!

ROBBIE: Tell us more about your time in the Pentagon.

MAX: The Pentagon institutions we dealt with mostly were ARPA (Advanced Research Project Agency) and IDA (Institute for Defense Analysis), as well as the service organizations. IDA was the subcontracting institution, really a way to get around the limits of Civil Service compensation —Herb York was the big cheese up there.

MAX: We didn't have a good idea what the Atlas performance was, up until I got on that committee.

PAUL: We didn't have a need to know.

MAX: We had only implicit, not explicit, information on what its performance was. Implicit information said it could not get into orbit.

Our initial design used the second stage of the Polaris—a solid rocket. It had the Atlas carrying the Polaris second stage and then the Mercury on top of that.

KEN: This was the early Mercury?

MAX: Yes, the early Mercury. But we didn't call it Mercury—we just called it "the capsule." Abe Silverstein's big contribution was that he named it Mercury *(laughs)*—his only contribution to the program that I can recall. *(laughs)*

PAUL: There was a nuclear propulsion stage.

GUY: I spent a lot of time over at the AEC and the Pentagon with General Kearns, Jack Armstrong, Howie Schmidt, and Stan Gunn, and was briefed about that.

MAX: There were interesting things going on in that manned spacecraft meeting of ours. The Navy had a program—

PAUL: There was a Navy proposal, an Army proposal, and there was an Air Force proposal.

KEN: There were all three?

PAUL: There were all three.

MAX: And in addition to that, the San Antonio aerospace medicine people had a proposal *(laughs)* that kind of got shoved aside.

KEN: NASA wasn't formed at this point?

MAX: No, we're still in the NACA.

Ken: So the NACA had a proposal.

Max: Well, we didn't really have a proposal. We had a *design.* *(laughs)* Our group was set up with the idea of deciding how this thing was going to be built, and who would build it. But the Air Force refused to admit that Eisenhower knew what he was talking about. *(laughter)*

They really were gung-ho that *they* were going to do this. They would say, "Well, you know, we –the Ballistic Missile Division (BMD)—was then running three programs—Thors, Atlases and Titans. It turned out the BMD managed the Thor. They used Ramo-Woolridge to do their R & D. Ramo-Woolridge (later TRW) had supported them in the early design and development of the Titan. Well, the Titan was still at least a year or a year-and-a-half from first launch. The Air Force couldn't imagine using that, and they hated the idea of using the Atlas.

There was another one that was proposed to the Air Force—a metal cloth parachute that had a capsule that was nothing but a pressurized sphere. They put this thing in orbit on an Atlas. And when they got ready to come down, they'd deploy the parachute. By changing the rigging of the parachute for variable drag, they could navigate the desired downrange location. *(laughs)* And it was going to weigh two thousand pounds. But there was a guy from General Dynamics, a man named Frank Dorr. So when we had a break, and I got Dorr out in the hall, and I said, "Frank, what's this about putting two thousand pounds into orbit with the Atlas?" He said, "Yeah, we call it the 'Bare Atlas.' It will put two thousand pounds into orbit."

Paul: That was the first time we ever knew information on what Atlas could do and what a spacecraft had to weigh.

Max: At that time, I knew we could do it without having to use a second stage. Previously, when we had a solid rocket upper stage, we also had an escape system in the event the Atlas failed. Now suddenly we had no escape rocket. Shortly after we decided to go with the Bare Atlas, Bob Gilruth said, "Max, what are you going to do when the Atlas blows up?" I said, "I don't know!" He said "Well, you'd better figure something out!" *(laughter)*

A guy named Woody Blanchard had invented this tow rocket. He had designed a system that had a rocket with two canted nozzles positioned on the far end of the launch arrangement. It had a towing cable about fifteen feet long fastened to an eye between the nozzles. The other end of the cable was fastened to the test vehicle, which was resting on the rear of the launcher. Both the tow rocket and the test vehicle were equipped with fins for individual stability. The test vehicle was released at burnout and provided aerodynamic data during subsequent coasting flight.

Guy: It was the cart before the horse. I had to redesign the rockets so the jets didn't impinge on the model, and also design the pyrotechnic separation system.

Max: When Bob gave me the escape problem, I thought, "Gee, all we have to do is use a tow rocket, but it has to be in place all the time to pull the capsule away from the launch vehicle.

Ken: That was their launch escape tower.

Max: That's where the launch escape tower came from — Woody's tow rocket. *(laughs)*

Paul: Anyhow, we were at a meeting in Secretary McElroy's Office—he was the Secretary of Defense—at a time when they were trying to decide what kind of heat shield they would use on their Redstone. They were planning on using an ablating material that would neither melt nor otherwise lose mass, and this guy named Kantrowitz was working on the big copper shell for the ICBM, and he spoke of a stainless steel cone that would take care of the rest of it. I said, "I don't believe it will." He said, "What do you mean you don't think it will?" I said, "Well, I got the 4,000 degree ceramic pebble-heated air jet we'd built."

After we showed some ablation materials, before and after, they asked for the idea of stainless steel. So we put stainless steel in there and it looked like a flash bulb, it burned so fast. And that is why we both thought stainless steel was a very bad idea. Marshal used ablative material for the Redstone. And Wernher von Braun was sitting in the back yawning.

MAX: Those guys in the Air Force were trying to make that Discoverer nose cone work—they were determined to make it out of copper, because copper has a high conductivity. Normally, heat would travel into the mass of the nose cone as opposed to just the surface. And at the onset heating rate of a ballistic missile coming in on a ballistic flight path, if the copper was more than two inches thick, the front end got just as hot. Anything beyond two inches, the conductivity wasn't great enough to keep the surface cool.

Anyway, the big name called Ramo-Woolridge had decided that the cool surface compared to a hot one provided a stabilizing effect to the boundary layer, which in turn would reduce the heat transfer rate. So, we flew highly polished nose models, using four different grades of diamond grit to polish them to an extremely shiny surface. You thought you were looking at a mirror! It still overheated compared with their theories. We had their guys come out and look at it before launch. We very carefully kept a piece of plastic over it that was pulled away about five seconds before liftoff so it wouldn't have any fly tracks on it or anything. *(laughs)* That was part of our education: theory is great but don't depend on it unless you have experimental correlation!

ROBBIE: So, during these NACA meetings on the sixth floor, Max was hanging out with ARPA and Guy was running around the Pentagon going to meetings.

GUY: Yes, I was going through the Pentagon with the Propulsion Science with the Nuclear Rocket Program, and with other areas working on big liquid rocket engines and dealing with various things. I already knew everything going on in the Solid Rocket world. Sometimes I went to a meeting with Max if it seemed like there might be some propulsion item. We'd choose. We'd get a list of the meetings that were available for us to go to and what the subjects were. Zimmermann, I assumed, got those from somebody. I don't know how we were cleared to go over there. I don't remember. All I know is I showed up and they let me in.

MAX: I think I had a badge.

PAUL: Dryden arranged to have our people have access to all of these meetings on the basis that we weren't trying to steal anything from them, we were trying to help them. Our job always had been to help the customer on the part of the U.S. government.

GUY: My feeling was that I was trying to steal everything; I was trying to bring everything in that was related to space!

PAUL: We were trying to learn what their people were up to so we could get an appreciation for the problems that they were facing and know what we could tell them about what we knew based on their needs, not just what happened to be of interest to us.

MAX: But to end up—when NASA was created we had to have a space program. That was the final, the ultimate goal. What is the space program for next year going to be all about? You have manned space flight, you have unmanned space flight, you can look at the planets and a bunch of things like that, and we also had to have a technology program. That was it! And it had to be created by our group. When it was all over, Dryden divided NASA up into two groups, one under Silverstein for Space Flight, and one under Crowley for Aeronautics, airplane flight. Aeronautics and space were then two separate entities.

ROBBIE: So while Max was going to the ARPA meetings, and Guy was at the Pentagon, what were you doing during these final NACA days, Paul?

Paul: I was doing anything that I was asked to do. One of the things was to justify building the Goddard Space Flight Center on Department of Agriculture grounds out in Greenbelt, Maryland. I didn't think we needed it, but Dryden had been working to get NACA to become NASA, and like a good trooper I said, "I don't believe in it, but I will write you a good story about why we ought to have it." So I did, and Dryden approved it.

Guy: Didn't Dryden have some contact within the department?

Paul: And I think one of the Lewis guys and I had to do a real rough and ready story about *if* we *should* someday go to the Moon, what would be the value of the stuff that we could bring back? Was there anything there that might be intrinsically valuable if we brought it back? And we could concluded that even if it were paved with cut diamonds, it would probably not be a commercial success for the first flight. *(laughter)* It might someday be one, but we wouldn't be able to justify space flight on a direct economic basis.

So, we needed to figure out other good reasons why we ought to be doing this. I went to some of the same kinds of meetings that these guys did. I had built up some credence with the groups, likely because I rescued them from Kantowitz's suggestion that they just use the stainless steel cone instead of ablation on the Redstone. I was talking to them about what they knew, what they thought a space program should do, what we could do in the space program that could help them? As I recall, back in those days, whatever seemed to be a pressing problem that day, I would try to do something about it.

Aerospace Medicine, Animal Research, and the Form-Fitting Couch

Max: About a month after Sputnik, the Russians sent a dog named Laika up in a spacecraft. They didn't bring the dog down but it stayed up and stayed alive for about a week, I think. They were taking measurements on it. Our aerospace medicine people were just absolutely going crazy *(laughs)*. They were jealous. The Russian doctors were getting all this information and *they* weren't getting anything. So, they had their own project.

Paul: One of the so-called "spook" agencies in Washington came down asking the guys at the PARD what kind of data you might get about how animals and people would react in space, but they couldn't tell us where they got the questions they were asking. They couldn't tell us what agency they were from, and it was none of our business, because our business was to provide information to anybody that the government thought ought to have it. That's what we did, all our professional lives—any authorized government person could come in and pick our brains for anything. We didn't try to hide anything. We answered every question as best we could, based on our knowledge and our understanding of the question.

And it wasn't until a couple of years later that I finally figured out who these people were. They were monitoring all the Russian telemetry work and trying to figure out what these wiggly lines meant. They didn't have the foggiest notion what they meant!

Max: They were getting flight data and probably decoding them. But all they were getting was squiggly lines, and they didn't know what the measurements were, because they didn't have the calibration, they didn't even know what it was hooked up to. That was kind of interesting.

This was before NASA. There was this period up until March where we didn't know who was going to run the space business for the United States. The aerospace medicine people in San Antonio really wanted to get in on this. They had a large colony of primates to study the effects of the environment of flight such as radiation, acceleration, and atmospheric pressure on these animals. And they also wanted to spend some of their research dollars on space. They had a desire to do some experimental flights. We had a desire to test a vehicle that they could use on such flights. We were kind of discussing things together to see if we could team up. But Gilruth was scared to death of these doctors.

Robbie: Why was Gilruth scared of the doctors?

Max: Well, they were saying we've gotta fly so many monkeys and so many chimpanzees before we put a man into orbit. You might recall the first [Russian] cosmonaut went up about a month before Shepard did. Well, at that time when he went up, we were in a big argument with the medical people and the National Academy of Science on how many chimpanzee flights we ought to make. They wanted us to make *ten* chimpanzee orbital flights before anybody could go up. We said no, we would make *one*, and if it was successful, we would go ahead. They wanted ten because one was not "a good sample." Statistically, they didn't believe in "one."

Robbie: Neither is ten! *(laughter)*

Max: That's right! We said, "We are going to put a guy up in sub-orbital flight for six minutes of weightlessness, and we think we'll have a pretty good idea of things after that."

Paul: They said they wouldn't require the ten chimp flights if we would *kill* 100 chimps on the centrifuge.

Robbie: You mean kill them while you were testing them?

Paul: These were professional MD, PhD researchers, accomplished scientists who would compromise from ten chimp flights down to "just" killing 100 chimps on the centrifuge.

Robbie: Why did they want to do that?

Paul: To learn more about the effects of these high accelerations as well as the flight take off.

ROBBIE: So they just wanted to keep stepping up Gs until they died, 100 times.

PAUL: Of course with that blastoff, they needn't. For one, that would have decimated the world's supply of chimps. We didn't *have* that many. Plus, we didn't have the time to make 100 centrifuge runs.

MAX: We were doing things that no one had done before. One of the things we got criticized about right away—we knew that during entry, the crew would take an acceleration of 8 Gs, plus or minus a half-a-G, something like 8 Gs, for an extended period of time. Now, people had taken 10 Gs for very short periods of time and it was great to see how well they could do.

I got a hold of a bunch of reports that the Germans had written during World War II for centrifuge tests on their fighter pilots. They were ahead of us in flying jet airplanes and rocket airplanes. They actually had some with rocket power. They could say, "Yeah, these can go fast but they will be taking a hell of a lot of Gs if they try to maneuver them." They were trying to figure out how to do that. They had run some tests with people in the supine and prone position for an extended period of time. I forget the exact data, but they indicated a really good probability that in the supine position, which means on your back, that you could take 8 Gs for a good period of time. At that time, we came up with the idea for the form-fitting couch. If you look at the body and put it under 8 Gs, you say, "That's not much different than whipped cream at 8 Gs. It's going to sag."

PAUL: We also used a couch with the body immersed in water. And Max said, "Well, how much water do you need?" He decided finally that just the water content of your epidermis really is enough. So, he built a form-fitting couch.

MAX: Then we had to test it! Along about this time, the Navy had given up the idea of being the agency for flying people in space. But not the

Air Force—some of our most vigorous critics now became the Air Force medical people. They were very concerned that they were not going to be in charge of this project, and they were criticizing everything we did. Fortunately, it was the Navy that had the best centrifuge—at Johnsville. So we sent a test pilot, Bob Champine and his form-fitting couch, to Johnsville, along with one of our young engineers to go out there. A man named Heberlig.

MAX: Heberlig helped me with the couch in figuring out how to make it form-fitting. We'd made plaster-of-paris casts of Champine in the supine position, and then used that female to make a male plug and then a couch that form-fit the plug. Champine rode it up to 8Gs. We had the profile, which was pretty much a duplicate of what you're experiencing in entry. It was just a straight line ramp up to an 8 G level period of about fifteen seconds, and a straight line ramp down.

ROBBIE: Was this after the pig was killed?

MAX: No, this is early on, before the pig. This is when we were trying to make sure the couch would work during *flight*, not during a landing. There was this guy named Carter Collins who was part of the centrifuge staff, who was a bit slimmer than Champine. He'd been doing all sorts of tests to increase human tolerance to sustained levels of high Gs. They actually had a metal suit in a sitting position that they would flood with water, with a test subject inside. This minimized the effect of the Gs on the test subject, since the body has the same density as water.

I don't know how they were going to put that in an airplane. *(laughter)* That really worked pretty well, so Collins said, "Gee, you know, I would like to try that couch out." So he found some Styrofoam and some other kind of hard foam, and he made some wedges so he could use Champine's couch. Every time Champine would take it up to a high G, Collins would take it to the same G. So when they got it up to 8 Gs, Champine said, "Well, I did my job." He was ready to go home!

PAUL: He headed off to Edwards.

MAX: *(laughing)* He was not going to be a damn astronaut, he was a test pilot. A professional test pilot.

KEN: Those were the days when astronauts weren't all the big heroes.

MAX: However, Carter Collins said, "Well, let me see if we can go a little higher." So he went up to ten, and then he went up to twelve. Management out there wouldn't let him go but two Gs higher every time. So he went to twelve Gs, and Jack called me up about that time and he said, "You know, Carter Collins has gone up to twelve Gs."

Meanwhile we'd run a few extra trajectories. We asked, what happens if the Atlas fails at a velocity near 18 thousand feet per second? At that velocity, the centripetal force from circling the earth at constant altitude would only be about one-half G. So you would start falling and end up entering at about a ten-to-twelve degree flight path angle. A ballistic capsule would then experience about 18 and one-half G's deceleration from drag. So, I told Jack, "If he is willing to ride it up to 18 and a half to twenty Gs, I'd be forever grateful." The next day, in two G increments, he rode up to 14, then 16, 18, and 20 Gs.

ROBBIE: Did it hurt him? Was he injured?

MAX: He was not injured but he was some kind of sick! *(laughter)* He was pretty well pooped out. Your inner ear gets affected with these high Gs. Your inner ear is what tells you what's up and what's down. It does with three semi-circular fluid filled canals that provide a function similar to a set of gyros. Each canal has some little hairs that sense flow of the fluid when the head is rotated. Your sense of balance is primarily related to the output of the inner ear. Carter's inner ears were badly traumatized by the centrifuge from the combination of excessive forces on the inner ear sensors and the Coriolis effect. Of course he was holding his head really still. What happened is that his inner ear got extremely overloaded and lost its sensitivity. He left the centrifuge around four or five o'clock right in the middle of traffic. He had to drive, I believe, fifteen miles.

MAX: He said he had to stop three times to vomit on the way home

KEN: So he got the early motion sickness.

ROBBIE: But there was no permanent damage to his ears?

MAX: No, no permanent damage to anything. He was in great shape. The Air Force was upset. You know what their attitude was? "Those guys in that Navy centrifuge facility are taking terrible risks!" *(laughter)*

When the time came to test the couch for the Gs the astronaut would experience during landing, Heberlig designed a couch for a pig, who was to be hoisted to a height of 16 feet and then dropped into a sandbox, where it would land at a velocity of 30 feet per second—the predicted rate for Mercury. But, strapped into the couch, the pig died before the first test could begin. Heberlig was baffled, and for a while there, it looked as if humans might never get up off the planet because they couldn't come back down! But finally the farmer who had sold Heberlig the pig explained that pigs always die if they are kept on their backs for any length of time, as their organs press on the lungs and the pig suffocates. As a result of the farmer's explanation, the period between strapping in and launch time was drastically shortened, and the tests were a success.

Formal Education, or the Lack of It, at Langley

PAUL: There was this argument about whether NACA should become NASA, or should the Air Force or somebody else run the space program? The National Academy of Sciences had a little study panel to evaluate the various places. Since Langley was kind of the top NACA Center, being the oldest, and we thought the smartest—of course, others didn't really agree with that—anyhow, this panel decided to do an evaluation of Langley to see just how qualified we were to run a big program. They came down and in just a few minutes they decided we were totally unqualified. Do you realize that of the 3500 employees that we had at that time, we had only three "professionals"?

ROBBIE: According to whom?

PAUL: According to this National Academy of Sciences panel!

ROBBIE: And how were they defining "professional"? A college degree?

PAUL: An earned PhD! We had three on the staff who had earned PhDs in some branch of science.

ROBBIE: All of you had college degrees but not PhDs?

PAUL: Well, some of us didn't even have college degrees. The best design guy, Caldwell Johnson, had two years of college—

GUY: In the Langley Apprentice School.

PAUL: In terms of brain power, he should outrank anybody!

GUY: You have to realize in those days, very few people even got a Master's degree. Most people, particularly engineers, did not go on to the PhD. If you were an engineer and you got through your Bachelor's degree, you were going to go out and work and make money. Engineers were not in it for academic achievement—they wanted to get the job done! I was afraid if I went to graduate school, when I got back, some young kid who had a Bachelor's degree would be sitting in my desk holding down my job.

PAUL: In those days, per diem was still close to six dollars a day. That was for room and board and everything.

The Creation of the Space Task Group

ROBBIE: How did the Space Task Group get created? You guys were all up there in DC.

MAX: Well that's a *good* story all in itself. I was with Bob Gilruth about October 10th, or something like that. NASA got born on October 1st, 1958. We had been working on this manned capsule program ever since early spring. We'd already done a lot of tests. We'd done some escape tests using some rockets that Guy came up with. We'd done parachute tests. We'd done landing tests to check our landing bag—we'd done a lot of stuff. It was all kind of basic research—it was charged to the research budget of the NACA. So we went up to talk to Glennan [the first NASA Administrator], who had come in about a month before—he came in around the first of September.

PAUL: This meeting was not held until October 4th or 5th.

MAX: Yeah, it was the first official meeting on manned space flight. So we were briefing Glennan, who hadn't had a briefing on the program. Bob and I went through what we'd done and where we were and what we were planning to do. When we got finished, there was a long silence. Finally, Glennan said, "Well, what comes next?" Bob looked at him and said, "Well, you know I've got to have authority to go ahead." Glennan said, "Oh! OK, you've got the authority." Bob knew he was supposed to be running the program, but he hadn't been assigned a staff or offices and other facilities that would be needed. He just had authority to do it. He had a budget—he had a $30 million budget.

PAUL: Bob told me that the actual words that Glennan used—the first three words almost gave him heart failure. They were, "Get the hell—" and Bob was sure he was going to say, "out of here." But the next four words were "on with the project!" *(laughter)* That's how it got authorized.

MAX: Bob went back and talked to Tommy Thompson and said, "You know, I have got to have some people." So Tommy said, "Well, let's form the Space Task Group."

ROBBIE: Where did he get the name, why did he think of it as "the Space Task Group"?

MAX: It was a silly name! We had to have a name—it was going to be a *temporary* name. Tommy Thompson actually created the organization. He directed me to organize the Space Task Group. See, it's not in capitals here. [*Looks at the actual memo (see below), laughs.*] That was kind of presumptive because the administrator never talked to Tommy Thompson, but he said to implement a manned space project. "This task group will be located at Langley Field," and so forth and so on. Finally in the second paragraph he puts the capitals on it—Space Task Group.

[Here we provide a reproduction of the original memo]

NASA – Langley

November 5, 1958

MEMORANDUM for all concerned
Subject: Space Task Group

1. Effective this date, a Space Task Group reporting directly to NASA Headquarters is established at Langley Field, Virginia to implement a manned satellite project. Mr. Robert J. Gilruth has been appointed as Project Manager and Mr. Charles J. Donlan as Assistant Project Manager.

2. The following Langley Research Center employees are hereby relieved of their present duties and assigned to the Space Task Group:

Bland, William Jr. (PARD)
Bond, Aleck C. (PARD)
Chilton, Robert G. (Flight Research)
Donlan, Charles J. (Off. Assoc. Dir.)
Faget, Maxime A. (PARD)
Fields, Edison M. (PARD)
Gilruth, Robert R. (Off. Assoc. Dir.)
Hammack, Jerome B. (Flight Res.)
Hatley, Shirley J. (Off. Serv.)
Heberlig, Jack C. (PARD)
Hicks, Claiborne R., Jr. (PARD)
Kehlet, Alan B. (PARD)
Kolenkiewicz, Ronald (PARD)
Kraft, Christopher C. (Flight Res.)
Kyle, Howard C. (Instr. Res.)
Lauten, William T., Jr. (Dyn Loads)
Lee, John B. (PARD)
Livesay, Norma (Off. Serv.)
Lowe, Nancy C. (Off. Serv.)
MacDougall, George F., Jr. (Stab.Res.)
Magin, Betsy F. (PARD)
Mathews, Charles W. (Flight Res.)
Mayer, John P. (Flight Res.)
Muhly, William C. (Planning Office)
Purser, Paul E. (PARD)
Patterson, Herbert G. (PARD)
Ricker, Harry H., Jr. (Instr. Res.)
Robert, Frank C. (PARD)
Rollins, Joseph J. (Off. Serv.)
Sartor, Ronelda F. (Fiscal)
Stearn, Jacquelyn B. (Off. Serv.)
Taylor, Paul D. (Full-Scale Res.)
Watkins, Julia R. (PARD)
Watkins, Shirley P. (Off. Serv.)
Zimmerman, Charles H. (Stab. Res.)

[signature]
Floyd L. Thompson, Acting Director

TMB.KOH

Copies to:
Director
Associate Director
Assistant Directors
Chief, Technical Services
Chief, Administrative Services
Assistant Chief, Tech. Services
Assistant Chief, Adm. Services
Division Chiefs
Branch Heads
Section Heads
Unit Heads
Each affected employee
NASA Headquarters
Files (3)

MAX: But anyway, Gilruth negotiated these people from Thompson.

PAUL: Tommy had decided to release the people. He was still outrunning Gilruth at that time.

MAX: Man, you talk about lifting yourself up by your bootstraps, that's exactly what happened! It was a very informal thing.

ROBBIE: What happened next?

MAX: When we got word that NACA was going to be NASA, Chuck Matthews joined us. Chuck asked me what to do. I said, "Well, we ought to start working on getting a contract." So he got a group together to write the specifications, and that included people like Caldwell Johnson and others. See, I was going to Washington every week with these guys, and Chuck was back there doing the specifications. This was before it was the Space Task Group.

ROBBIE: OK, because once the Space Task Group got established you guys didn't keep going up to Washington any more, did you?

MAX: No, no.

ROBBIE: Because then you had to run the program.

GUY: We still went up to Washington a lot to testify before Congress.

MAX: Yes, but not anymore to plan the agency. By that time, it was up to Dryden to tell Glennan what to do, I guess. *(laughs)* The fat was in the fire!

ROBBIE: So when you guys got created as the Space Task Group, was your first agenda to implement a manned space flight project?

MAX: Oh, yeah. And we concentrated our full effort on that. We actually had a Request for Proposals out by late November. NASA had only been in existence since October the first.

ROBBIE: This was November 1st when the group was created.

MAX: By late November we were out trying to contract for it with the specifications.

ROBBIE: How did you do it so fast?

MAX: Because we didn't have a bureaucracy—you'd just go do it! *(laughter)*

GUY: Nobody was saying you couldn't do anything!

PAUL: All of us had kind of learned over the years that it's a helluva lot easier to figure out *how* we can do it instead of to think of all the reasons why we *can't* do it. So when they said "Do it," we could! That's when we had the most fun—when we were *doing it*.

MAX: One of the good quotes now is "Just Do It!" You know, we'd been "just doing it" for a long while! We had a lot of practice just doing it.

ROBBIE: Why did you feel such urgency to work such long hours? Was it just that it was fun or was it Sputnik, or what drove you?

MAX: Both! We knew that the Russians were trying to put a man up there, and we didn't want them going up first.

GUY: That carried all the way through the Apollo program. Everybody I had working for me was willing to work whatever amount of time was necessary with no overtime pay.

MAX: Everybody accrued a lot of leave, and finally they put a cap on the leave you could accrue.

GUY: All you had to do was drop the word that there was something that needed to be done. You never had any trouble getting the people to do what was needed. Everyone was dedicated and challenged and interested to whatever was to the limit of our abilities. We were all in our mid-to-upper 30s, and full of vim and vigor, and were enthused about everything. The other thing was that we didn't have any bosses who put any impediments in our way. That was one of the real keys to working in that operation. It was a total team effort where everybody could participate to the limit of their abilities. Big Joe.

MAX: Along about two months after NACA became NASA, and the program was officially authorized, I was in California at the Ballistic Missile Division trying to get an Atlas D, and one guy said, "You know, we think we're going to have two surplus Atlas Cs, and we don't know what we're gonna do with them." I couldn't help but ask—you know, we'd been through hell with the damn guys at Redstone trying to get them to give us a decent schedule. So, I said, "When do you plan to use those?" He said, "We're going to fly them if we have a use for them—we'll probably fly those next August." Now this was in November when we learned they would fly them next August! So I said, "OK, we'll take those two and we'll use them to launch some full-scale reentry heating models." These were what were often referred to as "boiler-plate" vehicles since the structure was heavier, and they were equipped with a minimum of operating systems.

KEN: That's how Big Joe got started?

MAX: Now wait a minute. At that time we had the Mercury Request for Proposals out but had not yet chosen the contractor. We didn't know what the final shape of the vehicle was going to be. We put out a work statement which said, "We think it ought to be this shape or something like it." Of course McDonnell Douglas comes in and they changed the lines a little bit. They made the size change a little bit. We got them under contract in the middle of January. The minute we got a letter of contract with them, we had a shape, a final shape. *(laughs)*

We sent half the work for the boiler-plates to Lewis. They agreed to do part of it, and Langley was going to build part of it. Lewis was going to put the attitude control system in it. They sent two of their two attitude guys to Langley. We used a compressed air attitude control system. And we didn't get the damn thing built in time. We actually got through about two weeks late. *(laughter)*

KEN: That's remarkable, given that time period!

MAX: As I said before, the vehicle was designed so that in the event that the attitude control system failed, we would still make reentry. And we *expected* that attitude control system to fail. Now on the very first flight that we made, which was Big Joe, we had a full-scale external aerodynamics exactly like the Mercury with a parachute in it and an attitude control system. We planned a trajectory that would reenter down range, closely duplicating a return from orbit. So we put the thing on the Atlas and launched it. The Atlas failed. It failed to stage.

KEN: This was the first time you ever put anything on top of a booster like that, wasn't it?

MAX: Yes, it was about the third time the Air Force ever flew an Atlas—*successfully*. They had a bunch of launches where it blew up, but this one didn't blow up. But the engines failed to stage off. It had three engines, two big ones, and at—oh, I guess about 120 seconds or something like that—the two big engines are shut down and the flight is continued with the center engine. The big ones are supposed to be jettisoned. But the big engines did not drop off.

So, the Atlas keeps going with the unaccounted for weight of these engines. The Atlas was supposed to reach a velocity of something like

24,000 feet per second, at which time it would shut down all propulsion. But because of the extra weight of the two big engines when it got up to about 18,000 feet a second, it ran out of propellant. It never turned off the propellant valves. So when combustion ceased, some propellant and the gases that pressurized the tank flowed through the nozzle and continued to produce a low level of thrust. This was enough to continue the production of micro-acceleration. *(laughter)* So, it was going along probably to two-one hundredths of a G.

KEN: It was still struggling!

MAX: Big Joe had a backup separation trigger. It was timed to implement separation five seconds after the predicted time for the cutoff signal from the Atlas. Of course, Atlas never transmitted the cutoff signal. So the thing goes along until finally the backup timer on Big Joe comes in. This releases the Marman clamps that were used to attach Big Joe to the Atlas and it's ready to separate. The Atlas was equipped with retro-rockets to help effect payload separation. These were to be fired by the cutoff signal, which never came. And the Atlas didn't back off, and there was still sufficient thrust coming from tank gases exiting through the nozzle to keep the Atlas plugged into Big Joe. Meanwhile, about ten seconds after expected separation, the Big Joe attitude control system is implemented by timer. It was instructed to make Big Joe do a 180-degree flip. So, it burned up all its compressed air trying to turn the *whole Atlas* around. So no attitude control system! But, it made a good entry! *(laughs)*

KEN: It did? So you still got the entry information in spite of a non-working attitude control system.

MAX: And we did have a failure on the first launch. The aerodynamic shape was such that it weather-cocked. We had one instrument in there that was very interesting. It was a sound detector. Nothing but a microphone and a tape recorder, is all it was. We had that on tape and we could hear the entry. It would go whoosh, whoosh. The whooshes would get louder and faster. Whoosh, whoosh, whoosh, whoosh, whoosh, whoosh, whoosh, whoosh, whoosh [*increasing tempo*].

As the Gs went up, the frequency went up and the amplitude decreased. This is the same effect you get when you strum a banjo or you get a rubber band and you flick it. You get one frequency, and if you pull it tight, you get a much higher frequency. Frequency goes up, amplitude goes down. Well that's exactly what we could hear going on during entry! *(laughter)*

PAUL: He had figured this out sometime before, because when people would think that the thing doesn't have any aerodynamic stability, it does in that flight plan because as it's coming in, the air gets denser and denser to do this deal of tightening the string.

MAX: We'd done that analysis before. That wasn't an unexpected thing. We got confirmation on the whole thing right there. You couldn't have asked for a better flight. *(laughs)* The only bad part was that the pickup ship was supposed to pick up, I think, at about 700 to 1000 miles further down range.

ROBBIE: Guy, did you participate in Mercury even though you weren't part of the STG?

GUY: I did all the rockets—the launch escape rockets and the retro rockets, I was the guy who went and worked with McDonnell Douglas on this part of the program.

MAX: Before that, we had to design a rocket for the escape rocket. Of course, we needed something to put in the specs, so we had a "for-instance" rocket, which was nothing but a paper design. But we also needed an early rocket for some early escape tests. For that we had to modify an existing rocket to a three-nozzle affair, so that the jets would not be directed right at the capsule. Guy was able to obtain the existing rockets—Guy got them to change the nozzle. We finally made our first flight about time the NASA was authorized.

GUY: When we were evaluating the proposals to build the Mercury capsules, we used a numerical rating system to rank the contractors. I was in charge of the propulsion review team and gave one contractor a zero. Bob Gilruth called me in and said "Guy, you shouldn't give anyone a zero. Why did you do it?" I said, "Because I couldn't give them a minus." They knew the escape rockets resultant thrust should be along the capsules axis, so they had all the nozzles pointed down right at the capsule so they passed through the capsule's center of gravity!

Von Braun, Little Joe, and the Issue of Control

MAX: Von Braun was a neat guy. He was a real character. He'd charm the pants right off of you.

PAUL: He had built-in loyalty from this group. He was always Herr Professor. He was not "Dr. von Braun," he was "Herr Professor." In Germany that was a much better title.

MAX: The whole center used von Braun as a figurehead—he was very charming and of course he could open any door for them. His people came over in this so-called Operation Paper Clip—they got themselves out of Germany before the Russians got them. At first we put him in New Mexico, at White Sands and Alamagordo. Then they moved him to Huntsville. They were concerned about their heritage. So von Braun was not only their boss, but he was their social leader, and the Germans really adored that man. He deserved it, because he was more than just a good rocket scientist—he was a hell of a leader. He had all of the top qualities of a good leader. I dealt with him both as a companion and also at one time as a competitor during the first part of Mercury.

ROBBIE: What were you competing with him for?

MAX: Well, he really wanted to take charge.

ROBBIE: Von Braun wanted to take charge of the manned space flight program?

MAX: When I was with the ARPA people, they had their own program, which was to put a man on top of a Redstone rocket in a Jupiter nose cone. This nose cone was not an ordinary cone—it was cone-shaped with a small-radius round nose on it of ablative material. Think of it as a solid. They drilled a hole right through it about 30 inches in diameter, right through it. And they put a capsule, a 30-inch diameter capsule, that just fit inside this nose cone. The capsule was only about 4 feet long. They had this one German named Jack Kuettner, who was gonna ride this thing. He was a test pilot for the V-1 buzz bomb. Kuettner was small, even shorter than I am. And he was going to be von Braun's test subject. It would make its normal vertical flight and then release its nose cone to reenter the atmosphere, and when it got down to about 10,000 feet, it would blow this little capsule out the side, the parachute would open, and the capsule would be lowered to the ground.

ROBBIE: This thing never flew, I take it.

MAX: No, he didn't fly this mission, but it indicated that von Braun was very interested in it. Of course, this was the kind of thing *our* test pilots held their nose up at—this idea of being a medical specimen. When we got the authority to go ahead with Mercury, Bob Gilruth said "You know, we've got to have a test plan," We dealt with Otto Winzen, who had the high altitude balloons that could carry the capsule up to something like 60,000 or 80,000 feet and expose it to a vacuum for like a day—24 hours—and drop it for recovery on a parachute. Then the next thing was to put it on top of a Redstone and essentially make the same flight they were planning with the Jupiter nose cone. I thought, gee, you know, they fly these Redstones all the time—this is going to be something cheap and easy to do. We went down to negotiate the costs of the Redstone, and it cost almost as much as the Atlas. They had all kinds of overhead they put on the effort. And furthermore, after we got all signed up for the Redstone, we more or less were told, "Deal with von Braun because he's the head of the Redstone project."

They said, "When are you going to send the capsule down here to Huntsville?" We said, "We aren't going to send the capsule. Aren't you going to launch it down at the Cape?" "Oh, yes, but it has to be integrated with the Redstone." "Well, what do you mean?" "Oh, well, we've got to make sure that it doesn't have any electronic interference with the Redstone." They had this great big metal cage, a Faraday cage is essentially what it was. And inside external electromagnetic waves interfere with the integrated test.

PAUL: Radio waves.

MAX: Yeah, radio waves. They were going to talk to the Redstone and they were going to tell it "We're going to talk to the Mercury." And they had this great big program. And we had a deuce of a time trying to get that thing within reason. I think we finally did have to send one down there, but we really shortened the program.

Meanwhile, we wanted some short flights to test out our escape tower, and also to make some other short flights with monkeys. So, Paul Purser and I came up the idea of what we called "Little Joe." Little Joe from Bonanza was named after the same thing as Little Joe the rocket ship. Throw a pair of twos, that's called a "Little Joe."

GUY: Four the hard way is "Little Joe" in crap shooters' language. Four the hard way is two and two, you see.

MAX: Well, we had four rockets on it. And the four rockets were all bundled together—1, 2, 3, 4—in a bundle. Then we put a container around them, just an aerodynamic shell. Rockets burn their propellant very rapidly and we didn't want that high acceleration, so we fired two and then we fired the other two. So it was two plus two—Little

Joe! I think we bought all of our Little Joes for the price of one of von Braun's Redstones. We contracted with Langley for the construction and launches at Wallops Island. By that time, the Space Task Group was a separate entity.

GUY: I was up at a meeting of the American Rocket Society in New York one time, and I talked with this guy from Chrysler—von Braun's brother Magnus worked for Chrysler—telling him that you [the Space Task Group] would really like to have had the Jupiters rather than the Thors, and the reason that the Thor was winning out was that the Jupiter just cost too much. So, man, von Braun comes onto me like I don't know what. He wanted to know who I am and what the hell is going on. I told him, "You are just playing too damn hard to get. If you are really interested in doing this, you are going to have to do something about getting the price down to where it is competitive with the other people bidding." *(laughs).*

GUY: There was always a lot of competition between the two Centers. I don't know. I think maybe because the management between the two Centers really didn't see eye to eye on a lot of things.

MAX: When the whole thing started coming apart was when we got into the lunar program. We needed a very big rocket to do the lunar program. We had a rocket that had eight engines on it—the Nova. Then it came down to fish or cut bait. Wernher said, "Five engines is all I can go." But that wasn't big enough, so we had to use either a lunar orbit rendezvous or an earth orbit rendezvous. He wanted to use an earth orbit rendezvous, which kind of, again, would put the whole program much more under their control than ours.

ROBBIE: So why didn't you want it to be under von Braun's control? Why didn't you just want to let him have it and work with him instead?

MAX: We trusted ourselves more than we trusted anybody! *(laughs)*

We were just that arrogant.

Robbie: Why wasn't von Braun just part of your team?

Max: We'd been having this trouble with him with the Mercury, trying to get agreements with him. He was somewhat hard to deal with, on occasion.

Paul: He was very convinced of the validity of his own thing. When you could talk to him quietly and explain, if you made sense, OK, and I was able to get along with him very well, but then Walt Williams came along, and he and von Braun drew sparks off of each other like mad. He was hard to deal with, even with good friends! Von Braun was *almost* at the stage of really being a member of the team, until Walt started working on it.

Lunar Orbit Rendezvous

Max: The problem with earth orbit rendezvous was that it required two launches to get to the Moon. It required an awful lot of time in earth orbit before you could deploy. There were just a lot of question marks, a lot more risks. You might say there was less human risk in the earth orbit rendezvous than in lunar orbit rendezvous because we were going to have to make a rendezvous at the Moon, and all the people involved would be right there. The number of abort options was not well understood then. It turned out that we had a lot of abort options. But, getting the program done in a decade was damn tough, at least we thought so, and I still think it would have not have been so tough had the fire on Apollo One not occurred.

We started off with the idea of what you would call "all-up lunar flight"—launching a huge rocket. It would land on the Moon, and the same vehicle that landed on the Moon would come back. And the vehicle that would land on the Moon would have to reenter the atmosphere. We started on the design of this vehicle, and the more we studied it, the more we realized that it was going to be a very, very complex animal.

It turned out that the only way we could land on the Moon was to use what we called a "lunar crasher." You could not take a single rocket and depart from earth on the single rocket and have that rocket land on the Moon, because the rocket, by the time you landed, would be such a huge thing that the return capsule would be way up high and it would be very difficult for the pilots to land this thing.

The main thing the lunar module had to do was land on the Moon. It was designed *just* for landing on the Moon, not for traveling out to the Moon, not for traveling back from the Moon, not for reentering the atmosphere, not for entry steering. We cut a very complex problem into two compound problems. Once we understood the implied difficulty of designing a pilotless vehicle arrangement that could both land and launch itself from the Moon, then reenter Earth's atmosphere and land by parachute, anything but lunar orbit rendezvous was unacceptable.

We had two people we had to convince *(laughs)*. We had to convince von Braun, and we had to convince Harrison Storms. North American wanted to have their vehicle land on the Moon. That's what they wanted the contract for—to transport the astronauts to the lunar surface. They would be willing to build the lunar module to fulfill their contract. We said "No, you are not going to land on the Moon. You are just going to provide the transport to lunar orbit."

ROBBIE: What about the name "Apollo," where did that come from?

MAX: That came from our self appointed namer.

ROBBIE: Why name it for the Sun God when you were going to the Moon?

MAX: I don't know! It came up on Silverstein's list and he called it Apollo. He felt that it was his privilege to name these things, so he called it Apollo.

Visions for the Future: The Rules of Risk

ROBBIE: What is your vision for the future of humans in space? If you could paint the picture any way you wanted, what's your vision?

MAX: I think the long range vision is that humans will explore the inner planets. I'm not sure we'll go much beyond Mars, unless we get hyperdrive or something like that and the human race is then going to move to another solar system. We need space for more and more things. The commercial use of space requires human applications.

I'll make another prediction that's not original with me, because Bob Gilruth said it long before I did. He said, "Next time we go to the Moon, the people who do it are really going to find out that it is very difficult." I would agree with him on that. With the modern ground rules, it's almost impossible to do. The military talks about "rules of engagement." Well, I talk about *rules of risk*. Those rules we have imposed on ourselves almost preclude the possibility of a landing on the Moon in the next thirty or forty years. So we either change the rules or we won't get there for another thirty or forty years.

It's the mindset of the bureaucrats on what's safe and what's not safe, and it's the culture of the country where human life is so precious that it can't be wasted. I'm talking as an engineer who is used to looking at big pictures and understanding what's going on.

ROBBIE: So you think risks have to be taken to get back to the Moon and they won't be willing to take those risks.

MAX: Every endeavor in the future has to be more risk-free than the previous ventures. You just have to look at history to understand that.

GUY: There's no joy in success without an opportunity to fail. If you can't fail, then there is no point in trying. If anybody can do it, it's not a challenge.

ROBBIE: So are you saying the astronaut that gets in the next rocket has to be willing to die?

MAX: I didn't say you'd have to be willing to die, but you have to understand the risk. The risk right now of flying is acceptable because it's an existing machine. But if you designed another one, you'd have to preclude a lot of the risks that are acceptable on the shuttle—they'd have to be ruled out.

GUY: Many of the astronauts live around here—Buzz Aldrin, David Scott, Dick Gordon, Gene Cernan and Mike Collins all went to the Moon. Jim McDivitt, Rusty Schweickert and Walt Cunningham flew on pre-lunar missions as well as Gemini missions. My wife knew all of their wives. She said, "Well, what's the difference in working in NASA?" And every wife told her, "We go to fewer funerals these days" because they were all test pilots, and the death rate of test pilots is far more severe than any other group in the United States.

ROBBIE: How do you calculate the true value of human life? You said that they are overvaluing human life right now and that's hurting them.

MAX: Well, that's the problem. It depends on what you're talking about. The place where a human life is valued the least is in the highway department. I can go up and down this road here and show you situation after situation where humans are at unnecessary risk because of the way the highway is constructed. They're slowly getting rid of these things, but I think the highway department probably used a formula—a million dollars for a human life, something like that. If the person is a criminal, their life is probably worth about $20 million. You spend $20 million dollars with him after he gets the death sentence. We send them to the gas chamber in five or ten years. It doesn't make any sense.

GUY: I've worked in safety. I've worked in the most hazardous part of the operation. Everything I had could explode and detonate and do all sorts of damage, and all of the failures of anything I had to do with were going to be the most spectacular things you ever saw. Consequently, I was very heavily involved in safety. My attitude toward safety is that *if in order to save one man's life, I have to spend more than one man's lifetime trying to do that, there's a net loss to humanity.* That's a very cold-hearted way of looking at it. It's a very practical way of looking at it. That would not be a very popular opinion, but I've always looked at safety that way.

MAX: The best way to be safe is simply to understand what the hazards are. If you understand what the hazards are, and the people that are involved understand what the hazards are, then you've taken a big step forward toward safety. A lot of our safety now is what you might call "plastered-on safety." It hasn't got anything to do with making the thing safe, but we've got a committee to sit in review, we've got special organizations that specialize in safety.

The organizations that specialize in safety don't understand the problem as well as the guy that is working with the problem. You know—you've got novelists that write good novels and you've got critics. The critics themselves would love to be a good writer but they don't know how to do it, so they become a critic. Same thing with movies. They don't know how to act or they don't know how to produce a movie, so "What the hell, I'd like to be a critic." It's cheap to be a critic. Well, you've got these safety organizations of nothing but critics. That's what they are. Most of those people have never really done real engineering themselves. They had a brush with it or they've got an engineering degree, but they don't know how to make things work, they can only criticize the work of others. *(Laughs)*

ROBBIE: As you speak, I am remembering that if you had been listened to in the design of the Challenger, there would have been no O-rings.

MAX: The rocket would have been all one piece. It wouldn't have had any joints.

ROBBIE: So that means it wouldn't have blown up.

Max: But that was not a politically acceptable solution.

Robbie: Right, I understand. Guy explained that it should have been built by the company near Cape Canaveral, which could have made it all one piece, but the rule was it had to be put up for bids, which meant it had to be in two sections just so it could be transported. So design simplicity is one of the ways to stay safe?

Max: Of course it's one of the ways to stay safe! If it's simple, now you know that the possibility of a hidden problem is very low. If it's complex, your probability of a hidden problem is very high. If the problem is hidden, I don't give a damn how many safety committees review it, if it's hidden to the people that knew and developed the system and worked on the project, who have the most intimate knowledge of what's going on, the chance that these outside reviewers will find it is probably one percent, compared to the probability that the actual guys working on it will find it. So they might add one percent improvement in the chance of finding the hidden problem, at best. And if you don't have any hidden problems, now you should have a reasonably safe system, assuming no terrorists or bombs.

Robbie: Max, you are famous for the simplicity of your designs. Can you pick a few examples and describe to me how you stuck to simplicity in the face of pressure to make the design more complicated?

Max: Well, the best example I can recall concerned the lunar program. The bidding contractors were told that only storable hypergolic propellants could be used for the vehicles that we were responsible for. This meant that the propulsion system would be much simpler and more reliable than those using cryogenic propellants.

However, there was a really reliable little rocket engine made by Pratt and Whitney, the RL-10. It used hydrogen and oxygen, cryogenic propellants. The warm hydrogen would then have enough energy to drive the turbine that powered the propellant pumps. Nothing ran very hot, and the engine could run for hours. Not only that, but it was simple to start and shut down. Naturally the Pratt and Whitney rep came to see me. I sat through a long lecture on the many virtues of the RL-10 and agreed with him. But an engine is only part of a propulsion system. To make the lunar mission, we anticipated a number of shutdowns and subsequent restarts. With cryogenics, you have to deal with boiloff, settling the propellants, pressurizing the tanks and measuring the used and remaining propellant. While some of these requirements would also apply to storable propellants, the methods used to deal with them are ever so much simpler and reliable. So we had to eschew the RL-10.

GUY: Luckily, Max didn't ask for my opinion on this or I would have straightened him out! With Apollo 13, we had this huge Board of Investigation—every Center director of NASA was on the board.

MAX: Are you comparing Apollo 13 and Challenger?

GUY: I'm talking about how we went about the investigation. We never got any real outside help. We did it all internally on Apollo 13. Even the people who were assigned to investigate it didn't play a really big part. *We solved our own problems.* Our attitude was, "Hey, we've had a problem. What we are going to do is not try to fix the blame. No one is going to be blamed for this. It obviously was strictly an accident, an honest accident. We want everybody to come forth and to tell us everything they know about what happened. No one is going to be punished and there is going to be no retribution. All we want to know is exactly what happened so we can go fix it." And we had absolutely no trouble. The people who were responsible for the accident became sort of heroes afterward for fixing their own problem. That's the way Apollo 13 worked out, anyhow.

With the Challenger, if someone in authority had jumped on that thing right away and not let it get out of hand, it would have been handled much better than it was. It would have never gotten as far as it did. They made all the mistakes they could possibly make, and we didn't do that.

In the end, I told George Low that we had to make two small changes, one to a switch and one to a procedure to fix the problem. He couldn't let us do something that simple because of the outside pressures to do something major, so we spent about $10,000,000 on an unnecessary redesign. Privately, he admitted I was right.

ROBBIE: Tibby, what is *your* vision for the future of humans in space?

GUY: If we want to do big things in space, one country is not going to do it alone; it's going to require total international cooperation. It's too big a task and too expensive for one country to do it alone.

Our future in space is the same as it always has been. The reason we're in space is that we're able to make observations from a totally different platform or viewpoint. You see, here I am on Earth, and all the instruments I have to make measurements are subject to gravity or various other constraints that our environment places on us in making observations and the ways we go about making observations. If we get free in space, we have a totally different viewpoint. We're free of a lot of things. And we'll be able to discover things by going into space that we are not capable of discovering here on Earth. I'm convinced of that. That's the real purpose of going into space. That we'll look at things, after we've been up there we'll see things from a totally different viewpoint.

All the things we are concerned about now might be discovered if we can get up there and look at a lot of things from a different viewpoint—we see that with the Hubble Telescope. The value of space is to look at things from a different viewpoint, just as Hubble is doing right now. There are other outcomes. Everything that goes into space has three important requirements. It must be small and light. It must work for its entire life unattended. And last, it must hardly use any power. That's why all your electronics and TV last for years without failure, or why you can use all these miniature electronic controls in every conceivable application even as small as watches, etc.

MAX: The astronomers on the ground get just as much information and can control what they're looking at through the Hubble Telescope, just as they do with the camera.

ROBBIE: Paul, what's your vision for the future of humans in space?

PAUL: I think we should work hard on the space station for all of the good reasons that Guy just mentioned, and I think we should either revisit the Moon or go to Mars. As to when and how, we're not going to do it until somebody outside the United States does something to shake up the establishment and the press, because that's what it takes to get the U.S. to move, is somebody outside the system.

For example, I remarked on how NACA grew during World War II and then how it grew again when it became NASA. But NACA started in 1915 when somebody else went to Europe to find out why they were building airplanes in Europe and not very many here, and he came back with a report that the Europeans had more military aircraft than we did. We had 14 military aircraft in this country in 1914 and '15. The Europeans had 1400, and they came back and reported that. And NACA was formed as an advisory committee to study the problems in flight with a view to their practical solution. And by the mid-30s, we were the leading air transport nation in the world, but we were still behind militarily because the German military-industrial complex was itching to take over. When Lindbergh came back and reported that, there again, we got up and did something about it.

CHAPTER

3

CALDWELL JOHNSON
SPACE CRAFTSMAN

BIOGRAPHIC INFORMATION

CALDWELL (pronounced "Cadwell") Johnson was born in Hampton, Virginia in 1919. His maternal grandfather was a pilot in the Virginia Pilots Association, which brought ships in through the capes into Hampton Rhodes. During World War I, they had a lock on that whole system, and every ship that went in, including naval vessels, had to pay these pilots, and they made tons of money. When World War II came along, the Navy said, "We've had enough of this," and they decided to train their own pilots. Many years later, the great battleship Missouri—the battleship that Japan surrendered on—ran aground right outside of Hampton Rhodes.

Caldwell grew up in a suburb of Hampton. It was a new community, and at least a half dozen of his neighbors were engineers who worked at the old NACA. Floyd Thompson lived around the corner. Hartley Soule lived a couple of blocks away, and Thomas Harris was there. His family socialized with these people. The family had beach parties once every couple of weeks in the summer, and he says he was star-struck by the guys who worked at NACA. He would hang around and listen to them, ask them questions. They seemed delighted to tell him things, so he "got an education of a sort, just being a kid eating hotdogs at a wienie roast."

When Caldwell was in high school in Hampton, the family faced hard times. Those were Depression days. People were out of work. His father lost his business, and even the guys working for NACA had to take pay cuts. There was a 20 percent across-the-board pay cut for everyone. But that was certainly a lot better than not having a job at all.

In the small communities they kind of made do one way or the other. One of his grandfathers, the one in the Pilots Association, had enough money to more or less tide over everyone else in the family. When it came time to go to school, his family had no money. Caldwell had an aunt who lived in Charlottesville, Virginia. The University of Virginia is located there and she offered to let him live there for free. He spent two years at the University of Virginia, but the family was out of money and it was hard on her trying to feed an extra mouth, and he went to work in the summer for NACA as a part-time employee. Thus began a distinguished career that included working on Mercury, Apollo, Skylab, and the Shuttle.

INTERVIEW

This interview took place on October 1, 1999 in Johnson's home in Clear Lake, Texas.

ROBBIE: When you were a kid, what were your hobbies? What were your favorite things to do?

CALDWELL: Well, we lived only a few blocks from Hampton Rhodes. There was a beach there and there was good crabbing, and at low tide you could sell crabs, and as kids we all had some kind of boat, so in the early years, life revolved around that waterfront. As I got a little older, with the influence of the NACA people, I started building model airplanes. It turned out that building model airplanes, essentially, led to what I did the rest of my life. I got the job because of this man I knew named Tom Haucher, who was interested in model airplanes. He'd organized a little model airplane club in Hampton. He was a very clever guy. Later on he worked at NACA for a long while.

One of our neighbors was Raymond Sharp, who later became manager of the Lewis lab in Cleveland. At the time that I went to work there, he was Chief Administrative Officer at Langley. He had been an aero-mechanic and had studied law at home, passed the bar exam and got a job in administration.

KEN: When you say "model airplanes," were you building them or were you building them and flying them?

CALDWELL: I was building them and flying them. Of course, they were not sophisticated like they are these days. Everything was just rubber-powered then. There were no engines or radios or anything like that. For a dollar, you could buy all the materials you needed. I built the things in my room. There were balsa wood shavings all over the place, and glue and things like that, and if you opened the door and there was any breeze, it blew it into the rest of the room. I suspect that my parents decided that I couldn't get into much trouble up in my room building model airplanes. I was about nine or ten when I started.

ROBBIE: And how long did this passion consume you?

CALDWELL: I still build model airplanes. As I said, it was probably the single thing that led me to do what I did for a career.

ROBBIE: Did you fly them at model airplane meets?

CALDWELL: Yes. Tom Haucher organized these events and provided transportation. He was one of the few people in the group who had an automobile, and he would take us to places like Norfolk and Richmond, and we would compete.

Raymond Sharp, the Administrative Officer at NACA, knew about this and wanted to hire me. He was a very dominant type of man, very direct. If he made up his mind and if he thought something was the thing to do, he just did it and didn't ask someone else. If he had to bend a few little things to get what he wanted, he bent them. The guy who ran that shop that they sent me to said he needed a kid to help him do some things and Sharp said, "Why don't you hire this fella Caldwell? He's just the kind of guy I want."

Well, before they could hire me as a permanent employee, they used the three dollars a day thing as a pretense to hire me. Of course, I didn't do any labor work. He sent me right to a little shop in the wind tunnel area and I helped to repair wind tunnel models. The first day I was there, they couldn't just immediately send me over to the shop; they had to make a pretense, so they sent me over to a building inspector and said, "You help him." I didn't know anything about building inspection. He said, "Well, go up on top of the full-scale tunnel. They're fixing the roof up there." They were using this asbestos-corrugated concrete kind of siding. You can't use it anymore because of the asbestos. He said, "Go up there." The tunnel was about 200 feet high and I was petrified at the idea of going up there, and this chief inspector seemed to sense the trouble and said, "Wait a minute. I've got another job for you." He sent me to another job. That day a man working there fell though the top of the full-scale tunnel and killed himself. That very day.

ROBBIE: That could have been the end of a great career before it got started!

CALDWELL: That's right. Anyway, I got along fine in that little shop with Tom. It was heaven. I would have paid *them* $15 a week just to let me work there, if I had it. I was working on model airplanes! Except this time it was real and you got paid to do it. It was in the spin tunnel where they test the spin characteristics of airplanes. It was just great.

Sprucewood Propellers

CALDWELL: Finally, a time came when I decided I was not going to go back to school, that I was going to seek to be a permanent employee. I asked about it and they said, "Well, you can but you have to pass a physical exam for civil service." Now that doesn't seem sensible—nobody does that now in the civil service, but that was the rule then.

They sent me to the Army hospital there at Langley. I went through the usual things, but I knew when they got to the color business that I was going to have a hard time, 'cause I'm red-green blind. I screwed it up terribly. You know these little color dot charts? I would read things that you shouldn't read and I couldn't read the things that you should, and that box of yarn with different colors of yarn. I was absolutely hopeless. I took the report back after the doctor had filled it all out, and he had marked something like "Failed" or "Not passed" or something. It was the end of the world.

Dr. Sharp was the Chief Administrative Officer, and his secretary was little lady named Miss Bloom. She didn't know what to do about it. No one ever failed! *[laughs]* She took it into his office and I could hear him bellow. He said, "Send him in here!" I thought, "Oh my God, I not only failed but he's gonna chew me out for failing." He said, "It says here you're color blind." I said, "Yes sir." He had a big map, on his wall, of the Western Hemisphere and he pointed to the Atlantic Ocean and said, "What color is that?" I said, "Blue."

ROBBIE: Could you see blue?

CALDWELL: Well, I knew it was blue. I could tell colors like that. He pointed to the middle of the thing, the land end, and I said, "Pink!" All maps are pink. Then he pointed at the Pacific and I said, "Blue," again. He said, "I've known you since you were a kid. I knew you weren't color blind." He took the doctor's report and scratched it all out till it just said, "Okay." He hired me. Can you imagine anybody in NASA now, if they got a report from a doctor, who would say, "Nonsense! I've known you since you were a kid," and just scratch it out? He went on to be the Director of the Lewis lab.

ROBBIE: How did you feel when he said you passed?

CALDWELL: It was like a whole new world starting again. I probably just flew back to the place. This was in 1938. It was obvious that there was gonna be all kinds of trouble. The war hadn't started, but Germany was doing things and the government decided to greatly expand NACA. People knew that something was gonna happen 'cause Germany was flexing its muscles in Czechoslovakia and things like that, so the NACA was building up.

Sharp said, "We're gonna need a lot of young trainees. We can get engineers and we can get journeyman mechanics, although they're in a great demand now, but we need some trainees. Where are we gonna get these young men to be trainees?"

ROBBIE: Trainees for what?

CALDWELL: Trainees to man the wind tunnels, to learn to be draftsmen, to learn to be machinists, to do this stuff.

ROBBIE: All the men were in demand to be soldiers.

CALDWELL: Everybody was kind of building up. They said, "Why don't we hire these model airplane builders? We've hired a couple of them and they've turned out to be just the kind of guys we want. They like to work with their hands; they're interested in aerodynamics; this whole field interests them. They're young fellas. They're just what we need."

Sharp went to the Civil Service Commission and he said, "I want you to create a job category for model airplane builders. I want you to create a register. We have to have certain qualifications: they have to be high school graduates; they have to have won a model airplane contest; and I guess they have to be able to walk." I'll be damned if he didn't convince the Civil Service Commission to create a whole category. It was called "Under Aircraft Model Building" and the pay was $1260 a year. Even with all this going on, there were punchy kids. They sent out notices all around the Eastern part of the country. I guess they figured there wasn't any use sending anything to the West Coast 'cause that was too far for somebody to come for $1260 a year.

We got more young fellas, all my age (plus or minus a year or so) from Chicago and Wheeling, West Virginia and Philadelphia and Brooklyn. The first thing you know, there were maybe a hundred or so under this civil service thing and they parceled them out all around the shops.

Jack Kinsler was one. In fact, if you go through the roster of people in what was called the "technical service" that ran the shops and the mechanical services; they were all peopled by these fellas now. Not *now*, 'cause most of them retired, but at the time I retired all these shops were run by the Kinslers and people like that—the same bunch that came there.

It was a stroke of genius that he thought of that. It wasn't only genius that he *thought* to do it, but that he *did* do it. I'm sure it's not an easy job to march up to the Civil Service Commission and say, "I want you to create a new job category."

One of the big jobs during this expansion was the building of all the new wind tunnels. They needed some big propellers, some 43 feet in diameter with 11 blades, with each blade weighing 1100 pounds. They were all made out of spruce, all glued together and then carved to the shape, virtually by hand. There were no shops around to do it, so Haucher said, "I'll take these model builders, and we'll build these things right here on the spot."

All the old mechanics said, "Ha, ha, ha, ha, haaa. You wait till those dumb kids get to start doing this, They'll cut their fingers off, they'll mess it all up."

Well, we built them. Tom designed the shop and the equipment and we built the damn things. He broke the job down into elements that everybody could work on. We were enthusiastic. We all were model builders. We all talked the same language. We all had the same hobby. We were all the same age. That was it!

Later on, he also created an apprentice program, but that was after this happened, because they realized that the sensible thing to do is to organize the training and not just let it be haphazard. Where the west area of Langley is now was just trees and swamp. The first thing they built was a shop to build these propellers. Originally, it was nothing but a big old shed. All of us boys worked there. Tom was the head man, and because I was the first one there, I got to be his assistant. You can imagine—you get 50 or 60 kids about 19 and 20 all working in the same big shop—you know there's gonna be some stories.

There was no sewage in the west area in those days. They'd built us a great big privy—a "two-holer" Government Issue privy. It was right there behind the shop. Well, one day as I came back along the little path back to the shop, I saw a wallet lying beside the path. Some guy, pulling his pants up or something, had dropped it. I picked it up and saw whose name it was and I took it back in and put it behind the desk. I then proceeded to holler, "Is anybody missing a wallet?"

One guy said, "My god, I don't have it!" I said, "I'm not sure, but it looked like something may be down in the privy." He said, "My god! I had $5 in it." He grabbed two long pieces of sprucewood—and made a pair of tongs and went out there "tonging." Everybody was laughing, and I took it and put it where somebody was bound to find it, and then I got a friend to tell him about it.

The War Years

Robbie: Did the architect teach you how to do drawings?

Caldwell: I had learned quite a bit in high school, but the guy who taught drafting there was the football coach. He couldn't draw a straight line, so he would just get us started and go out and do whatever football coaches did. Most of us were just self-taught in school.

ROBBIE: But they gave you the materials to work with.

CALDWELL: Yeah, and when I was at the University of Virginia in the engineering school. The professor there was writing a book on drafting and I was one of the better draftsmen in the class, so I was commissioned to do illustrations for the book. I got special tutoring because he wanted what I was doing to look good. This was before the days of computer-aided drafting, so everything had to be done by hand, with ink no less. The reason for the ink was that the reproduction facilities couldn't handle pencil so well. You could draw with ink on linen, which was quite translucent and you could make blueprints from that. Paper was not translucent enough and the lights were not bright enough, so you'd have to put ink on this linen.

ROBBIE: This was before xeroxing. When you reproduced a drawing this way, it would come out light on dark, instead of dark on light?

CALDWELL: Yes, and it was blue because whatever the chemical was at that time would make the background blue.

ROBBIE: So that's why they call them "blueprints."

CALDWELL: Yeah. People who worked around the blueprint rooms would get a lot of scrap stuff, and they learned that if you took it home and soaked it in nice soapy water, all the sizing would come off of this beautiful linen. It was the sheerest. nicest linen you could ever see, and they would make beautiful handkerchiefs and slips. That's the sort of thing that happened in the so-called "olden days" that the younger people don't know anything about.

ROBBIE: We were laughing about how people think "cutting" and "pasting" are just terms for deleting and inserting stuff on a computer. They don't know that we used to actually cut paper with scissors and paste it back together to move paragraphs around.

CALDWELL: That's right! So, from the engineering department, one day led to another and I got better and better and better at it.

ROBBIE: Was the military draft a problem? Did anybody try to draft you all for the war?

CALDWELL: Yes. Of course, we were all draft age. For most of us, the boards were deferred because the NACA would go to bat and say, "We need this guy more to work on airplanes and do research work than we do to get him shot." So, many of us stayed. We were inducted into the United States Army Air Corps and then just assigned to go back and work at Langley. After the war, we were all discharged. In the meantime, we didn't get paid by the Army. You would never know the difference. There was a downside to that. The obvious upside is you don't get shot. The downside is that many of us felt very awkward. Other kids you knew in high school were getting sent off and having their heads blown off and that sort of thing and there you are, sitting in Hampton, Virginia, getting paid. On the other hand, it would have been silly to shut down the agency and send everybody off, because they needed the airplanes.

ROBBIE: Did you have the sense that you were helping the war effort by designing better planes?

CALDWELL: Well, I certainly tried to make myself think that. There were a lot of things that directly helped the war effort. NACA was doing virtually nothing in those years except helping the military. We were making the aircraft fly faster, making them stronger, and just greatly improving them.

ROBBIE: You were testing balsa wood airplanes in the spin tunnel?

CALDWELL: Yes, but then they got more and more sophisticated, and pretty soon they were not all balsa by any means. They would put quite sophisticated mechanisms in them that would make them act like they had a pilot in them. They would do things that you told them to do.

Another big problem we worked on was that planes would ditch in the English Channel if they ran out of gas. But they were not designed to land in the water. Like most airplanes, they would behave very badly when they hit the water. We had a big towing basin at Langley that was made originally to design seaplane hulls. Hardly anyone used seaplanes anymore, so we turned it into a ditching facility. We'd try different shapes of airplanes and build the model dynamically and structurally similar and land it in the water and see how it behaved. Sometimes it was too late to do anything about the airplane, but at least you could put a placard in the cockpit that would tell the pilot the best way to land the thing. Sometimes you might say, "Usually the best thing to do is keep your nose up, but sometimes the best thing is to keep your nose *down* in that particular airplane—you're less likely to kill yourself if you do."

Robbie: How big were these model airplanes you were building?

Caldwell: Six or seven feet.

Robbie: How did you go from model airplanes to learning all about the engines and the details of the real airplanes?

Caldwell: Somehow, between a half-dozen people doing something, somebody would know about it. Besides, there were experts on everything at the Langley laboratory. If I had a problem, there were people I could ask.

Robbie: Did you ever spend time in real airplanes?

Caldwell: Very seldom.

Robbie: So you learned it all on a small scale and yet you came to understand the engines completely?

Caldwell: Well yes, but no one person knows all these things. You get advice from other people and they get you headed down the right track.

Robbie: You develop a discriminating sense.

Caldwell: Yes. I think I've always been inclined to be doubtful. I don't give a damn who tells you something. Gilruth himself could tell me something and I wouldn't necessarily say, "Oh, that's a bunch of shit," but in the back of my mind I'd keep the door open just a little bit and try some checks and balances to see if there was something just absolutely stupid about the whole thing. Actually, I shouldn't say, "Gilruth," because as far as I know he never said anything stupid to me.

Transonic Flight Testing and the Beginning of the Space Program

Caldwell: Somewhere in 1945, near the end of the war, a small group of us were pulled out of the engineering department and sent over to a place in the west area to design an air-to-ground missile.

Ken: Okay, here we begin the space program part.

Caldwell: Yes, though there wasn't any PARD [Pilotless Aircraft Research Division] yet.

Robbie: And whose idea was it to do this missile?

Caldwell: I think it was the Air Force and I think it started out to be an air to air.

Somehow in the course of things, Langley got asked to do something about it and they got a little AD hoc team. I think Hartley Soule was in charge of that. There was also a fellow named Bob Jones who worked for Soule. They decided to design and build it in-house, and a fellow named Harold Maxwell and me and two other guys were put in a little office. They took some aero engineers from the research department and we designed the thing. They built four or five of them.

They never did perform as well as we hoped. We simply did not have the infrastructure that it takes to do something like that. Various components were not advanced far enough. The idea was too big for the tools we had. They flew but they didn't perform and they would never have caught up with an airplane. They were damn lucky they got off the ground.

The airplane would have outrun the missile and it would have run out of fuel before it got there, if it ever found it in the first place. The guidance system was poor and underpowered. Still, it was a start, and we learned to do something without some great program depending on the outcome. You could afford to fail without it being a disaster, which was good because if the first thing you try out *is* a disaster, it tends to put a damper on it.

Langley had a Research Department, a Technical Services Department, and an Administration Department. The Research was the reason all the rest of us were there, because they determined the product of the whole thing. We provided engineering in the shops and built things and launched them and instrumented them in support of the Research Department. I was in the Technical Services Department but I didn't have an engineering rating yet. In other words, I was not a professional engineer. There was a kind of underlying, subtle caste system. The Research Department, man for man, ranked higher than Technical Service, man for man. It wasn't overdone and it never caused any trouble. In fact, I think it put a certain amount of discipline in the whole activity that was like the services. You've got to have a certain amount of discipline. You can't let everybody run everything. Somebody has to be at the top and somebody has to be second and somebody third and so forth, and there was just enough of that to keep everything honest.

Robbie: You're saying that it was a hierarchy, but it was a functional hierarchy?

CALDWELL: Yes, it worked. And yet we all knew that if push came to shove, the Research Department was gonna win. That was an accepted way to do it and it didn't bother anybody.

For instance, there was a thing called the Supervisors' Club. It was only the supervisors in Technical Services. It was like the non-commissioned officers' club in the service. It had tremendous power because anybody in the Research Department, unless he was a stupid fool, knew he didn't want to cross the "sergeants." That's what made the place run. There was an understanding between everybody. On the other hand, the Supervisors' Club people were not about to do anything that we knew was a stupid thing to do.

ROBBIE: Were you in the Supervisors' Club?

CALDWELL: Oh, yeah. It started long before I got there, years before. The man that ran the Technical Services Department was respected by all and feared by many.

ROBBIE: Who was that?

CALDWELL: His name was Ernest Johnson, no kin to me. He was the hardest-nosed, smartest guy—absolutely no way to fool him. He knew everything that went on. He hardly ever said a word, but everybody knew what they were supposed to do.

Back in the early days, I was working on the drawing board. At that time, all the charts that were used in presentations had to be created by hand. Sometimes they were photographed and turned into slides, but often they would just be cards with it all lettered on, like cue cards. I could letter well and was pretty fast at it, so often times when the little outfit that was supposed to do it would get overloaded, they'd press me into service. I didn't want to be a chart maker, but the boss said, "Do it!"

Well, one afternoon after everybody had gone home, I was up there trying to make some of these charts because the airplane was gonna leave that night to fly off to Washington to some big thing. Lo and behold,

Ernest comes wandering through the place—all by himself. You never knew when he was gonna show up. He came and looked it over and knew immediately what it was all about.

He said, "Boy, what are you doin'?" I told him and he said, "That what you want to do for the rest of your life?" I said, "No sir." He said, "I would think you might ruin that thing at the last minute." I was drawing with India ink. After he left, I turned the bottle of ink over on to it.

ROBBIE: On purpose?

CALDWELL: I got the message of what he said. He as much as told me to screw that thing up. He knew I was gonna catch hell and I knew it, but he also knew I'd never be asked to do another one. It was too late to make a new one. They had to go to Washington and do the best they could and you know, I never heard another word about it. After my boss first raised hell with me and all but had a hemorrhage, I'm sure he protested up the line and, somewhere along the line, somebody said, "Cool it." I never got asked to do another chart. It took a little bit of guts to do it.

In 1945, I was with that little group of people organized to build the "Mitchell." That soon expanded. About that time, we were getting some solid propellant ordnance rockets. The armed services had plenty of them by then and you could just get them for nothing. Gilruth had conceived the idea that this was a solution to get around the problems of the limitations of transonic speed. None of the wind tunnels would handle the transonic speed range. They could handle supersonic and subsonic, but the tunnels couldn't handle this transition period that airplanes had gotten to, known as the transonic.

ROBBIE: Is that why Gilruth built Wallops Island?

CALDWELL: Yeah. Gilruth worked in flight research with full-scale airplanes and he had conceived the idea to put test specimens on the upper surface of a wing where the velocity is higher than the airplane's velocity, and he could get transonic speeds there when the airplane was still subsonic. It was very limited because you could only do what you could put on it. They had gotten transonic speeds dropping bombs.

He had already been experimenting in that area with different techniques, so he said, "Why don't we take some of these solid rocket motors and put 'em on a little pilotless aircraft and we can go through the transonic range at any speed we want? There's no safety involved; there's no pilot; there's no problem with the aircraft itself—we can make it any damn way we want. We've got the rockets and we've got people who know how to build the instruments that we'll need. We'll need a place to launch."

The tracking radars had already been developed for other reasons in the war—the Doppler radars and the tracking radars—so you have all the elements you need to do pilotless aircraft research.

KEN: Is that where Thibodaux came in? On the solid rockets?

CALDWELL: Yeah, when we first started, Thibodaux hadn't come back from the war yet, but it couldn't have been long afterward that he and Max came along because in 1944, things were winding down in the war.

KEN: But that's where the "P" came from in PARD

CALDWELL: Yeah—Pilotless Aircraft Research Division. The first thing you know they'd bought or leased land at Wallops Island and they were putting little huts up there.

KEN: You didn't have the tunnel in those days.

CALDWELL: Didn't have the tunnel and it was kind of messy to get there, even after you got across, because the island itself has maybe a mile of salt marshes in between. The only house on the place was a lodge for a sportsmen's club from Philadelphia. They hunted geese and ducks over there and had leased this island all for themselves and had their own lodge on it. The Langley people just ran them out.

One of the rockets that we used was a British rocket that used cordite as the propellant. They were not very efficient, but they were free. We were over there getting ready to do something one evening, and we were in this little shack that was as big as this room. It had a little oil stove going over in the corner and we had these cordite rockets piled up in this shed with a damn oil burner going.

ROBBIE: In the shed with the rockets right there?

CALDWELL: Well, they really are quite inert. Cordite is kind of like dynamite—it takes something like a blasting cap. It won't go off, but just the same, it was kind of silly.

They were these round steel cylinders, and something happened. One of them in the bottom of the pile came out and they started coming out going "plung, plung, plung, plung," rolling across the floor. There were maybe five of us in that little shed and the door was about two feet wide. It was almost like one of those Keystone Cops comedies, but boy we got out of there!

I didn't go over there very often. My job really didn't have anything to do with the launching. I would only go to see if something that had been designed and built was okay.

KEN: At that time had you resolved this issue between being in the research side versus being in the technical?

CALDWELL: No. I was still in Technical Services and assigned there. At the PARD they transferred a whole section from Technical Service Division and I wound up in a section called Dynamic Model Engineering Section. They transferred that to PARD but it was "on loan." Technical Services just flat refused to give up the people. They said, "You can borrow them, but they still report to us." Now, they didn't exercise any real power.

Robbie: Did you actively seek to be part of PARD? Was that what you wanted?

Caldwell: I would have preferred to do that, but I didn't see that it made any difference. Then somewhere along about that time, I think, my grade and classification got changed. I wanted to do the pilotless aircraft research.

Robbie: Why?

Caldwell: Well, it was an exciting new field. It was PARD against the battleship admirals. All the wind tunnel people were the old battleship admirals. It was just like in the Air Force.

Robbie: What do you mean?

Caldwell: Well, it was this upstart Gilruth with this whole new technique threatening all these old battleships, the wind tunnels. To a man, they circled the wagons. What a great thing to be in this outfit, the Indians attacking!

Robbie: So in other words, Gilruth could see that the limits of the technology had been reached and that if airplanes were going to take the next jump, they had to go to these pilotless tests to move into the transonic realm, and you couldn't do that in a wind tunnel, you *had* to use rockets.

Caldwell: Yes, but don't put the wind tunnel people down too much. They never did completely solve the transonic thing, but they ended up with supersonic tunnels. Gilruth provided the means to get across, at that time, what was just a wall blocking everybody. He didn't try to break it—he just went around it.

Robbie: So they continued on their trajectory, developing more sophisticated tunnels, where he jumped around that whole paradigm and went toward a whole new thing.

Caldwell: Yes. And this was with, say, a hundred people. There were these great massive steel things that had great propellers, which I'd helped build—big steel tunnels, thousands of tons, sitting around the place. Langley used to have a bi-annual inspection to which all of the industry and government agencies were invited, and they would show you what we had done in the last two years—all the reports that had been written and the different things, what our problems were and what we'd solved. It was a big technical open house. One of those years, after PARD got really rolling, I think there were sixty papers given. Fifty-eight of them were by PARD.

The PARD Grows Up

Robbie: When did you first meet Bob Gilruth?

Caldwell: He lived right on the beach and I lived near it. There were three of us in that area, so we got together in a car pool. Eventually though, we said, "We hate to do this, Bob, but you've got to find another car pool." His work habits were as irregular as ours were regular. He would be late in the morning and he could get away with it, whereas we couldn't quite so easily, and in the afternoon, God knows when he would be ready to come home.

He was a great guy to ride and around and talk with. He would give me a hard time with puzzles in physics, like things that would border on relativity. He was quite an innovator. Wonderful, kind guy. He lived on the beach with the whole Chesapeake Bay right in front of him and designed a sailboat with hydrofoils. He demonstrated that if wind conditions were right, it would work, but it was really a terrible place to be fiddling around with a hydrofoil because you can't sail it up on the beach—you have to slow down and sink and take the hydrofoils up and get to the beach.

Robbie: When you moved into the PARD, people were telling you what they wanted designed and you were designing it?

Caldwell: Yes. About that time, Thibodaux, Faget, and Stoney came down. Most of them did very well, but they were thoroughly capable guys to start with. Also, the right climate was there. Sometimes even the brightest guy in the world can't get anywhere if he's in a place that's already got all the goodies. As I told somebody once, "We licked all the icing off the top of the cake. Then we left the cake for some poor cuss, and he's gotta eat the cake."

Robbie: What are some of the examples of the things you were being asked to design?

Caldwell: PARD, or the aircraft industry, would have some kind of problem with stability or control, or something like that. They would come to NACA and present the problem, and NACA would decide whether or not to tackle it. They would request what was called an "RA"— a research authorization, from headquarters. That would be authorization to spend X number of dollars on a certain thing. Then, let's say it was a Boeing or a North American, or something. We'd say, "What's your problem? Let's talk about it, and let's see what we propose to do to solve it." They would get together and, once they had agreed in general terms, the project would be assigned to a Faget or a Stoney or one or two of the young engineers. They'd say, "Here's the broad picture. You work it up in more detail and stoke the thing and get a very rough idea."

Then the guy would come to me and say, "Here's generally what this will be. We want it to do this and we want it to do that. This is the kind of rocket motor it's going to take to get the velocities we need." I would rough it out a little bit more and then talk to the instrumentation guys, and we would make sure we had a launcher that would handle it and we had the boosters that we needed. Then, my outfit would make the engineering drawings to have it manufactured. In the meantime, the instrumentation people would do their thing. Sometimes, it would be a family of things. It would be basically the same vehicle, but you'd change the wings, for example.

It wasn't long before the guys in the instrumentation area and I figured that we should standardize some things. It was silly to treat every one of them like an AD hoc project. So we started getting certain types of things that would only require changing one number, say, to modify them. It saved a lot of money in the shops because you didn't have to build a zillion different kinds of things. In the meantime, the shops developed all kinds of slick techniques for doing things that they had never done before.

KEN: That's why you could do things in the shops that you couldn't go out and bid on. Nobody knew how to do it.

CALDWELL: That's right! Nobody knew how to do it. It was a trade back and forth. I'd say, "Shops, we need to do this," and they'd say, "We can't do *that* but we can do *this*." We would trade back and forth and, in the meantime, they would remember what I said we needed and they would work on that. At the same time, I would learn their limitations. They had the instrumentation people doing the same thing and we would all get together. At lunch time, whenever we ran together we would talk about these things. It was a closed circle of people and we traded ideas back and forth.

I was very disturbed when Johnson Manned Spacecraft Center began. They completely abandoned that concept and decided to get contractors to do everything. That spelled the end of all the young engineers. How are they going to learn anything if you go get some contractor to do it for you? The young engineers coming into PARD had hands-on experience, right from the very beginning. They were down in the shops; they helped put things together. They learned what you had to know to be a good aeronautical engineer or structural engineer, and Johnson doesn't do anything like that. They are trying to do it, as I understand, and there is a great effort to get back into that mode. The trouble is there are 30 years missing, so there's a big gap. It's now to the point that most of the supervisory people at Johnson have never seen a machine shop or wind tunnel or a pilotless aircraft. They know how to fly in airplanes and they know how to write minutes of meetings and go to design reviews.

There was one really downside of this system that we were working. PARD would come to me; I would go to the instrumentation people; we'd both go the shops. There was a little closed loop there. The downside of that was if there was some poor guy in PARD who we had already determined was a loser, he was a goner. I wasn't about to assign our good talent to a guy whose track record I knew was terrible. It worked just great for the guys who had made a good start.

Robbie: Ah! Now I understand what Faget meant when he said "the cream rose to the top."

Caldwell: Yeah, literally. It didn't just rise by itself, it was promoted.

Robbie: Because the guys that weren't that good went after a design to be implemented and they couldn't get the support.

Caldwell: That's right. It didn't take me long to figure which guys were winners and which ones weren't. Now that's a terrible thing, but it made for efficiency. Purser was one of the lead guys in PARD. One time, Purser said, "It takes too long between the time that we take a job to you and the time we get it up to Wallops Island. There's no reason in the world to take all this time. I'm just gonna show you." So he came down to me with a particular job one day and he had been all around to everybody and had gotten this priority A #1 from Youngblood in Instrumentation, and the shops and I had been told that it was A#1 priority.

Well, we knocked it out in about a third of the time that it generally takes. So Purser said, "I told you it was ridiculous to take so long on these things." He didn't know it, but I worked on it myself with my best guy, and we did that all over the whole place. *Of course* it didn't take long if you take all the best people that work with you, but Purser went around telling everybody he had invented a whole new procedure! In one month, I think that Wallops did thirty launches—an average of one a day. We had finally gotten into the hypersonic range, with five stages. We were getting about Mach 16 or 17.

Robbie: What were you learning?

Caldwell: We were learning about the very high velocity stability of things like missiles and we were doing some work on re-entry heating. When you get going very fast, there's a great deal of heat generated.

Robbie: And do they melt?

Caldwell: Yep, and we actually melted some things.

Robbie: The nose cone would melt?

Caldwell: Yes, and it would soon become blunt, but it would get to a degree of bluntness where it would no longer melt off. So, you could put two and two together and say, "Well, let's start out with it blunt."

Ken: This was the beginning of laying the base for how you could come down from orbit. This type of information was needed for that later development.

Caldwell: I remember with film one time that there was a blunt nose that had a dimple in it, as though one had shot a bb at it. I shoved my thumb in and made thumbprints, and at the bottom of the thumbprint it was quite cool. It was hot as hell all around the thumbprint but not at the bottom. Suddenly people said, "It's because the air is stagnant, going right on over it and it's staying cool."

ROBBIE: How was that information useful?

CALDWELL: Well, it turned up later with Mercury. We'd found out some of the things that had worried people about the heating, and the drag and what the after-body situation is. And if you have fins, what's going to happen to the fins? At first, all this information was a little piece here and a little piece there—then after a while, the matrix filled in.

ROBBIE: So you were getting that reentry phenomenon of coming back into the atmosphere, even though you weren't going into outer space.

CALDWELL: That's right. In some cases, the trajectory could be shaped so that it would go horizontal to a known density of air.

Bootlegging Mercury and Starting the Space Task Group

ROBBIE: Did you have any sense at all, at this time, that you were building the basis of a space program?

CALDWELL: No.

ROBBIE: Was anybody thinking about going into outer space?

CALDWELL: Yes, because everybody thinks about that. After you get to Mach 18, you're getting close to escape velocity. There's a certain velocity that enables you to escape the earth's gravity or you can go into orbit. Any physicist can calculate it. Everybody knew that kind of the thing. The question was "How do you do it and then, after you do it, how do you come back in without burning yourself up?" Right then, we didn't have a rocket big enough that could hold a man, but the know-how was there. It was just a matter of getting a more powerful rocket and making everything a bit bigger.

ROBBIE: What did you do about safety? What were the standards and how did you handle the issue?

CALDWELL: It took a while to develop proper standards. When we started, there weren't any standards, and it became obvious that you can't just let people do whatever they want. If your judgment gets impaired, you're inclined to take risks that later on you know were stupid, so they developed very strict rules. One day, everyone was eating lunch and one of the Little Joes over at Wallops Island—lo and behold, the thing launched itself!

KEN: How did that happen?

CALDWELL: I don't know, and to this day I don't know that anybody knows. Somewhere in the electrical system, something happened and that thing launched itself. It went sailing off and did its thing, except there was nobody there and no instrumentation going.

ROBBIE: So it staged and everything and nobody was measuring?

CALDWELL: That's right! It was kind of a boilerplate test vehicle.

ROBBIE: When these rockets were launched and were staging, how did you measure what they did?

CALDWELL: There was a radio transmitter and various devices that could measure temperature or movement at a distance and radio the signal back—its position in space, between the Doppler radar, which will measure velocity going away from you, then this radar that was called "584 radar" which will measure for your azimuth elevation. So if you know the elevation and azimuth and you know the velocity, you've got all the things you need to be able to tell what the track is. You not only know where it is, but you know how fast it's going.

ROBBIE: So, you're making all this technical progress at Wallops; you're learning all kinds of stuff about rockets, and by now it's 1948 or 1949? Where are we?

Caldwell: Well, once Wallops Island got started in 1946, over the next twelve years it just developed more and more sophisticated pilotless aircraft that worked. That's what occupied me for that time.

Robbie: And the point, since you weren't actually going for space, was to develop airplanes that could fly at these transonic speeds and to learn, from the behavior of the rockets, what would be the behavior of the airplanes.

Caldwell: You might argue that the heating, for instance, wasn't important to the airplane, but it turned out to be important, even to something like the Concorde, which is a long way from being a hypersonic airplane, and many military aircraft are now titanium because they just go so fast that they're getting hot.

Robbie: As you were gleaning this information, were people using it to build faster airplanes?

Caldwell: Yes, they certainly were. Funny things would happen at these speeds. Control surfaces would not act the way they acted at subsonic speeds. All the forces that regard the movement just got out of hand. There were some cases where the control effectiveness actually reversed itself so the pilots found themselves in an impossible situation. Here was an airplane that was not responding the way it was supposed to.

One of the first things we found out was that you'd better put some speed brakes on the plane so when it got in that shape the guy could slow down so that he *didn't* get in trouble. Now that's a dumb way to do things—you make the thing to go fast as hell and then put a brake on it to keep it from going so fast.

ROBBIE: So this wasn't just about airplanes, it was about ballistic missiles and weapons as well.

CALDWELL: It all got slanted toward the military applications, because during the war that's where the interest was.

KEN: And toward the end of the Second World War, the Germans were ahead of us with the V-2s.

CALDWELL: That's right. That was another situation where, even if it wasn't a national program, everybody could see the handwriting on the wall. You don't have to be the President of the United States to figure out what was going with the V-1 and the V-2—that it's just another step and we'd better do something. We didn't have anything to compete with that at all. Not a thing, absolutely nothing!

Thank goodness they were no better than we were. We could not have countered them. What did they end up doing? A guy would have to take a Spitfire or something and fly underneath the buzz bomb and put his wing under it and try to tip it over.

ROBBIE: Shoot them down?

CALDWELL: Well, I don't know whether they could shoot them down or not, but one of the maneuvers was to fly up beside the thing and tip it over. In fact if you shot it, the detonation might break your airplane, as far as I know.

Somewhere along about this time, in 1958 is when NACA became NASA. Mercury was on the drawing board, because that's dated June of 1958.

KEN: But when did you first get assigned to even begin Mercury?

CALDWELL: We were not "assigned" at all! We weren't hiding it, but there was no research authorization to put money there. I was charging my time to supervising an outfit that covered all these other pilotless aircraft things, and I'm sure Faget and all these other guys were doing that.

Robbie: So into the mid-1950s, you were doing more and more of this rocket works at Wallops.

Caldwell: Yes. Near the end of that period a few of us would be bootlegging time to work on Mercury.

Robbie: What started the discussion? Was it when Sputnik flew?

Caldwell: Probably. That may have intensified it, but once that five-stage pilotless aircraft was doing Mach 18, you're gonna start speculating about what you can do. It was obvious which way we were going—it was just a matter of getting the right ingredients together. Then, also, the ICBMs were getting the Atlases and the launch fields built. We didn't have any launch fields. All we had were these solid rocket motors that just wouldn't do it. We started building the Atlas and then the Jupiter, which was really basically the V-1. The Germans at Huntsville were developing that into an ICBM.

Ken: So it was kind of like there was one technology and development for solid rockets, but there was a liquid rocket booster that represented the WW II vintage that was also being developed.

Caldwell: Yes, that's right. Someone else was developing the basic lifter that you need, the basic propulsion unit, for entirely different reasons, for military reasons, doing it for an ICBM. That's the missing piece of the equation that all of a sudden falls into place. Up to then we were stymied because the solid rocket motors did things but they just flat wouldn't do this job. They're still in use. The shuttles use two of them, but by themselves they're never gonna do it. They do a good job of the initial lift off.

But the interesting thing is the fact that there was no research authorization, as far as I know. It was just simply bootlegged, and the people who knew about it just chose not to say anything. I don't suppose we spent a helluva lot of money, either. No one had to write one big check. It was almost all just swallowed up in our salaries. Our salaries were paid just filling out a time sheet every Friday and it got charged to something.

Robbie: So whatever you were doing, you were going to get paid.

Caldwell: And it was not a big enough item that it would attract enough attention and make a big glitch in the records.

Robbie: That's interesting—it wasn't a big enough item!

Caldwell: And I dare say that 90 percent of the important things that have ever been done have been done that way.

Robbie: Was it Max who first came to you and said, "Draw me this"?

Caldwell: It had to be Max—I don't think it would be anybody else.

Robbie: And do you remember that first time?

Caldwell: No, I suspect he didn't even actually come to me. I suspect we were both somewhere for some other reason, and that particular group of people was there and a discussion started on one subject and ended up on another subject.

Robbie: So there wasn't a moment when Faget walked in and said, "I want you to design a spaceship." It just was part of a conversation?

Caldwell: A group of us were aware of what the talk was and if, for instance Max had said, "Hey, Gilruth says that we oughta get started on this," I'm sure he wouldn't have had a piece of paper.

ROBBIE: So what happened?

CALDWELL: I had this outfit whose job was to synthesize these pilotless aircraft. We didn't select the components, or the motors to anything, but we took the elements and packaged them together and did the engineering work necessary to know that structurally they were all right and the shops could build them.

The Fagets and the Matthewses and the Chiltons all had their piece of the thing, but they would end up with me reducing that thing to a picture. Somebody would write that it's gonna be a blunt body and it's gonna be this big and it's gonna have this in it and it's gonna have that, but it's just a narrative. Sooner or later you've got to have a picture, because the narrative can mislead you—you'll think everything is great, but you'll find that you can't put it together as you thought. What I did was put the thing in terms of a picture. Now, once you get the picture, people can say, "That's not what I meant," or "We need more space than this," and then you start modifying the thing and juggling it around. If you're a good designer, you don't have to go through too many trials and errors, and if you know the people you're working with, you can come close to hitting it on the head the first time.

It also worked the other way around. They would say, "It's no use for us to try to tell him this, or that—he already knows what to do with all this. Don't worry about the structure; he'll put the right kind of structure in. Don't worry about the man fitting in there, he'll get the man in. Don't worry about how it's gonna attach to the launcher—that's just part of the business—he'll worry about that."

So, it's come around full circle from what we talked about before. It's the same people talking to each other all the time and not a total stranger that you have to go to, if for no other reason than that they were only 50 feet away. That's worth a lot. You don't have to write a letter or get on the toll line, you just walk downstairs and if the guy's not there, he'll be there after lunch so you wait a minute.

And it's a great thing to be able to do it "in house." The Langley Memorial Aeronautical Laboratory had configured itself to have all the capabilities we needed. It could do everything. The only thing it couldn't do was something so massive that it was just beyond our capability. For example, consider those great propellers I told you about—there was no way to build the hub at Langley. The hub weighed 16 tons and there was just no shop at Langley that could do it.

It was a wonderful place to start something—all those little pieces that you need. There were guys who knew about engines, and guys who knew about rockets, and guys who knew how to build things—the best aerodynamicists and physicists in the world.

ROBBIE: Were you doing more drawings for Mercury? What was the process like?"

CALDWELL: I don't think I did anything more on the operational version of Mercury, probably because it was in good enough shape to turn over to a contractor. I started working on development models. For instance, we had the Little Joe series.

ROBBIE: What was the point of Little Joe?

CALDWELL: It would look like Mercury but would be unmanned and suborbital. The purpose was to investigate the models for floatation and stability—whether they really were stable as they should be. These would be full-sized, too. There were wind tunnel models built, but the wind tunnel people had their own engineering department. Somewhere along in here, the Space Task Group was formed. It became enough of a thing that a new organization was formed. Now it became a legal activity but I still didn't work for PARD. They said, "We'll transfer you to the Space Task Group. Now you're gonna be one of the club." I wasn't all that enthusiastic. I didn't want to leave the program; I had a nice new home on the James River and fishing was good. It was home.

ROBBIE: And you knew that transfer to the Space Task Group was going to mean you were going to have to move?

CALDWELL: I knew it was gonna come sooner or later. It couldn't go on at Langley.

ROBBIE: Why?

CALDWELL: Because of the "battleship admirals." I don't mean they literally fought it, but the mindset in the place was on the wind tunnels.

ROBBIE: So this was going to have to have its own space.

CALDWELL: It had to go someplace else. You can't do a new thing in an old place. Anyway, I resisted it and we kept on going and I was still part of Engineering and Technical Services Department.

ROBBIE: Is your name on the memo that started the Space Task Group?

CALDWELL: No, I was not one of the first ones. It was a going thing before I finally transferred. I transferred because McDonnell had already gotten the letter contract to do Mercury.

You'd think he ran the government, instead of the government ran him, but Mr. McDonnell said, "I want you to send some of your people out here. I want to see what these civil servants look like before I get too involved with them." Apparently he hadn't had the best of dealings with government people, so Max and Bob Piland grabbed me and two other people and we flew up to St. Louis to see Mr. McDonnell. I was sitting between Piland and Faget and they were pounding on me to go ahead and transfer to the Space Task Group.

That night when we got to St. Louis, a tornado hit, and it ripped buildings down and tore the wing off our hotel. So the next morning we got there a little early, and Mr. McDonnell wasn't there and everybody was talking about the tornado. It had blown buildings down and raised hell in St. Louis. Exactly on the minute, Mr. McDonnell walked in with his secretary. He had a little dossier on each of us. He put the first one down, and opened it and said, "Mr. Faget," then he looked around and Max identified himself. But the first thing he said was, "I know this tornado last night is on everybody's mind and we're gonna devote three and a half minutes to talking about the tornado and then we'll get down to business."

Then he looked at his watch and I don't think we used even half a second after he said that. These dossiers said when you're born, where you went to school, years of service, all that sort of thing. Here we had to sit there with a contractor who's supposed to be working for the government, and he's grilling us to see if we're fit people for him to work for. That's the way he was, but his team did a good job. It was entirely different from North American, like day and night. Not necessarily the quality of the people, but the way the company did business, and Grumman was different, too.

A Box of Rocks, Beyond PARD, and into Space

CALDWELL: Let's go back to the boilerplates I was telling you about. The guys from Cleveland would fly down on a Monday afternoon and leave about noon Friday and go back to Cleveland. They never did build their homes or transfer until some of them came here to Houston. It made for a very awkward way to do business, with them designing some of the stuff and shuttling back and forth. What was worse was that they had had no experience whatever in light aircraft structures or rockets or pilotless aircraft, and I took one look at the things they had and I just couldn't stomach it.

We had the biggest battle over this and the politics of the situation were really awkward. You can't go to Lewis and say, "Your guys have got a poor design here." We finally arranged a sort of compromise and they agreed to change the most objectionable features, but it was still on the books as something *they* were doing.

I had a turf battle with one of those guys at Langley. There was limited office space and we were quarreling over who was gonna get the office space. Both of us worked for Jim Chamberlin at that time. Jim decided that he was gonna let us fight it out and he wasn't gonna step in and make a ruling. Well, one Friday afternoon, this guy got on the airplane and headed back to Cleveland and I told all my guys, "We're gonna move him out of his office while he's gone." We moved all of his furniture out into the hall, but we got to a big five-drawer file cabinet, and we couldn't budge it. We absolutely could not budge it! I said, "The son of a bitch! He's bolted it to the floor." I said, "Well, let's take the drawers out." It had a security bar and a padlock on it, and I said, "Break the lock." We broke the lock and it wasn't bolted to the floor—all the drawers were full of rocks! R-o-c-k-s!

KEN: What'd he do that for?

CALDWELL: He was a rockhound. He collected gem-bearing minerals. He was a mineralogist and he had filled that thing with quartzite and all these different things, and that's why we couldn't move it! We took one drawer at a time out and moved it out into the hall and put the drawers back in there. Never heard another word about it. He could have really chewed me out when he came back and got me in trouble for breaking the security bar, but he didn't have a leg to stand on. He never said a word; and Jim never said a word. He had to find a new office somewhere; I don't know where he found it.

There were fun moments. At least for me, it was fun!

KEN: Well now, did your assignments change dramatically once you became a member of the Space Task Group?

CALDWELL: Yes. Then somebody else had to run the old outfit that I left at Langley. The pilotless aircraft stuff went on, but by then, I didn't have anything to do with it. They went on to do the Scout missile.

KEN: Yeah. The Scout went on a long way. So at this point it was a dramatic change in what you had done.

CALDWELL: Oh yes, because, now I was 100 percent Mercury.

McDonnell was perfectly capable, as far as I was concerned, of taking this system. I concentrated on the boilerplates because McDonnell didn't want to do that and it was the kind of thing where it was too complicated to write contracts and change them a million times if you change a development vehicle. It was easier to do it in-house. That's where the Little Joe series came about.

We finally ended up taking McDonnell's first production article. It wasn't quite finished, but we took it and turned it into a boilerplate. We used the basic structure, and then did some other things with it. It was flown on the Atlas, but it got wrecked in a rainstorm and had all kinds of problems. Before that, we had a test article that was designed and built in-house. It was suborbital, and when the Atlas was supposed to shut down, it still had a small residual thrust which, ordinarily, would not have presented any problems. Atlas would have done its job, but it had just enough thrust that the weak little system that we had to separate the spacecraft from the bit was not enough to pull away from the Atlas. So Mercury used up all its control propellant trying to turn itself around to enter the atmosphere at a proper attitude, because it was going wrong end first. It huffed and puffed and used up all its gas and could not turn the Atlas around. Well, finally Atlas ran out of gas and gave up and by that time the boilerplate had used up its gas, so it started re-entering wrong end first. But it turned itself around with its inherent stability and was picked up later. It was really a delightful test, because it showed that the thing was inherently stable enough that, with or without a control system, it would turn around and properly enter the atmosphere.

KEN: So you got the test but it wasn't planned that way!

CALDWELL: Yes. And we knew it would come down somewhere around Hispanola. We wanted to put some sort of label on it in Spanish so that if the regular recovery people didn't find it and the fishermen out there did, they could read it. My secretary was Puerto Rican and I told her what I wanted it to say and she said, "I'd better take it home and let my husband write it because he can write better than I can and he'll know better what to put on there."

She came back in a couple of days with a translation of what I'd said, and somehow the word got to headquarters in Washington that this note was going to be on there. They had a hemorrhage! They said, "Do you mean that some secretary in some little two-bit office in Space Task Group presumes to write a note about what to do with this thing? Cease and desist! We will send you a proper placard to put on it."

They went to some language professor in Georgetown and sent this thing back and I gave it to my secretary because I didn't know what the hell it said, and she looked at it, kind of puzzled, and said, "Well, I'll take it to my husband. Maybe he'll understand it." Then she came back and told me he said very few people in that part of the world would be able to read the Spanish because it was classic Spanish and it just didn't mean much of anything to the people on those islands. Headquarters said to use it anyway. Fortunately, our destroyer found the thing and we didn't have to worry about it.

ROBBIE: Can you imagine someone from the islands seeing it fall from the sky, finding it, and wondering if it was an ancient form of Spanish that came from the gods?

CALDWELL: From a UFO or from the gods! By and large, though, Headquarters interfered very little in anything. There was no micro-management. In fact, on that boilerplate that we put together and McDonnell built, the first production model that was converted was sent over to the shops in the west area to have the outfitting finished by us.

I was put in charge of getting all that done. We had a roped-off area, not because there was anything so "secure" about it, but because people were always getting curious about things and getting in each other's way—guys wanting to see. I was there one day and I saw this guy standing by the ropes and looking at it. Finally he stepped over the ropes and came up to start looking and, eager beaver that I was, I walked over there and said, "Excuse me, do you have business here?" He said,

"Oh, excuse me, excuse me," and he got back over the ropes. One of the guys came over to me and said, 'Do you know who that was? That was George Low." He was very high in NASA administration at the time. I didn't know who he was and George, being the nice guy that was—some other guys would have said, "Well, you snot, get the hell out of here! I'm George Low." I figured that any day I'd get a telephone call saying, "This is the end of the world for you, boy."

The Great Preparation: Going to the Moon

CALDWELL: If you take a handful of pennies and throw them in the air, for the most part, they will come down half heads and half tails, but sometimes they come down mostly heads. That's what happened with the space program.

ROBBIE: They could just as easily have all come down all tails?

CALDWELL: Or not enough heads to have made it turn out so successfully. Let's say the Russians posed no threat. That would have taken the heat off. Or if two or three guys had their wives get sick and they decided not to come to Houston, or there was an automobile accident and some of them were killed. A lot of the success of the Apollo program hinged on some very informal little incidents, such as a conversation across a lunch table. But everything fell into place—*everything*! The country had the money, we'd won the war, everything was just right. And a new outfit had started with a fresh bunch of guys. Headquarters stayed out of it because NASA didn't *have* a headquarters of any consequence. You couldn't do it today because there'd be so many damn meetings and reports you'd have to do.

And we did it by ourselves. Look at the poor space station today. You not only have to deal with Headquarters, you have to deal with Congress and with the President and an election's coming up, the Russians. We are working with people who speak different languages and our partner goes broke. As I said, we licked all the icing off the cake.

ROBBIE: What was your role in Gemini?

CALDWELL: Almost nothing. Remember that we were designing Apollo before Alan Shepherd took his Mercury flight. Gemini came along between the two and when I finished up with Mercury, it was time to start with Apollo, so I missed Gemini almost altogether. Besides, Gemini was really just an extension of Mercury—the team was in place; they knew what they wanted to do and it wasn't like starting from scratch at all. So there were a bunch of us who had done our thing with Mercury.

ROBBIE: How did you feel when Mercury flew?

CALDWELL: Well, of course I was delighted, but I was so busy with Apollo by then. I won't say I didn't pay attention, but it was a long way beyond what I had any control over. When it gets to that stage of the game, I've done my thing a long time before. I was delighted that it went so well. I was part of the first group of people to come to Houston, not long after Shepherd's flight. Apollo was well on the way, as Mercury was, long before John F. Kennedy said, "We're going to the Moon." It was as far along as Mercury was before there was any research authorization for it.

ROBBIE: So you bootlegged Apollo, too?

CALDWELL: Well, it was surely known. The whole world knew somebody was gonna do it.

ROBBIE: Did you bootleg Apollo deliberately to go to the Moon? Was that the intention from the beginning?

CALDWELL: Oh yeah. There's no question of that.

The amazing thing about Apollo is that, as originally conceived, we had not really worked out the landing phase on the Moon. There was a controversy over lunar orbit rendezvous versus the direct approach.

Politics were involved, of course. In the first place, the people at Marshall, who were going to do the launch vehicles, were all for launching a satellite and rendezvousing in earth orbit. The Lewis lab had been promised they'd be able to build the big lander—the propulsion system that would go down on to the Moon, so they had their own ideas.

This, of course is a famous story that's been told many times. John Hoboult at Langley, who was not part of the Space Task Group, had calculated that, everything considered, the lunar orbit rendezvous was the way to do this thing. The first time you think about it, of course, is that we've hardly begun to orbit the earth and who ever heard of orbiting the Moon?—much less being able to launch yourself and join up with something already orbiting and head back to Earth.

It just seemed absolute nonsense, so the first reaction was that this theory sounds good on paper, but we can't do that. Well, it turned out that a bunch of us were working on this direct lander on the Moon. We couldn't do that either. We started doing the engineering work and it was impossible. You simply could not afford to put this whole thing down on the Moon and fly the whole thing back. In my mind, it had to be either John Hoboult's idea or nothing.

Anyway, Max Faget thought Hoboult's idea was absolute nonsense. By that time, I was in the Space Task Group and reporting to somebody else. So I said to Owen Maynard, "Let's go and talk to John Houbolt and see what he has to say about this thing." Well, the day that we were supposed to go over there, something happened and I couldn't go, so Owen went by himself and when he came back he said, "He knows what he's talking about. He really has got a good case."

About that time, Chuck Matthews got in touch with John and he convinced him that it wasn't a bad idea to rendezvous at the Moon, putting down a small craft and coming back up to the command module and then coming home. Suddenly, two or three more people decided they want in on that idea and they started running the numbers, and all the things that John said turned out to be a very good thing. Well, Max didn't yield very gracefully. In later years, Max implied that he knew it all the time, but I'll tell you that he was not very enthused about the thing at all.

Robbie: Max wanted to land the whole thing and bring it back?

Caldwell: Yes, he wanted to kind of brute-force it. But whatever you bring back, that whole thing has got to re-enter the atmosphere. The interesting part about this is that, in the meantime, the contract had been let to North American before this was decided, and the thing was being engineered, and North American, was, for all practical purposes, designing it as if it was going to be a lunar orbit rendezvous, even though no decision had been made. No one was working on this great vehicle that was to go down and yet, politically, North American did not like the idea of another contractor getting a piece of the pie.

Ken: Grumman.

Caldwell: They didn't know at the time that it was going to be Grumman, but they knew they weren't going to get this other contract too. So here we were, politically—they didn't like this lunar orbit rendezvous, but we engineering people were designing the command module as though it was going to stay up there in orbit. When we laid out the crew arrangement, we had the tunnel up there so you could dock, and you could get out of it or you could put the LEM up there. Structurally, it was designed to take all this kind of thing. I can't imagine a program being so screwed up and coming out of it so well. But in the back of people's minds, apparently, they were convinced of lunar orbit rendezvous—it's just that the program itself had not faced up to the politics.

Robbie: So everybody was essentially designing for lunar orbit rendezvous without an official policy. Is that what you would call a "self-organizing system"?

Ken: Yes, I would, because it was the right thing to do when you look at the other alternatives.

Caldwell: It turned out to be the right thing to do.

ROBBIE: What were *you* doing in all of this?

CALDWELL: Just what I always did—drawings. I did the first drawings of the command module. It had been decided that a three-man crew was the sensible size crew. In the first place, the more people you have, the more trouble there's likely to be. On the other hand, in a lunar orbit rendezvous, somebody better stay up with the craft, and it wouldn't hurt to have two going down to the surface, so the three guys worked out fine. It worked out just right for a lunar orbit rendezvous.

Well, everybody said, "Three guys are going be gone a lot of days, so that means you can stand an eight-hour watch, just like you do on shipboard." The fact is that we have never flown a mission where people stood three eight-hour watches. Everybody sleeps at the same time and everybody works at the same time. This business of having a man on watch was never used.

ROBBIE: Why?

CALDWELL: There's nothing to watch for! *[laughter]*

ROBBIE: You're just floating around in space.

CALDWELL: And the ground is monitoring everything.

KEN: There are no glaciers out there.

CALDWELL: There are no sand bars or pirates to board you! But some of the people argued that that was the reason for the three men, that each guy could stand an eight-hour watch. So that's another case where you do the right thing for the wrong reasons.

ROBBIE: It was you and Max Faget?

CALDWELL: Yes, then all of a sudden, I didn't work for Max. Also, he was very interested in this big group landing on the Moon. He wasn't for the lunar orbit rendezvous, so he was concentrating on that. He also had an ongoing responsibility for Mercury that he could not shuck away. He was committed, from an engineering standpoint, to support the Mercury program, so he couldn't just take off. But I could spend a hundred percent of my time on Apollo, and I did.

ROBBIE: What did that 100 percent of your time consist of? Every day when you went to work, to work on Apollo, what were you doing?

CALDWELL: Well, I made some very early sketches of the lunar module. I laugh when I look at them now—I wish they would get lost or something, they're so naive. Owen Maynard was a real crackerjack draftsman. He and I were talking about the landing gear for the lunar lander, and at that time there wasn't much known about the surface of the Moon. I think people like Owen and I had decided that it was just like Arizona or any other place on Earth where it never rains. It's got rocks and dirt and gravel and sand. Right at that time, two very prominent physicists, whose names I don't remember, said the place was just great pools of dust, maybe miles deep. Drop a stone on it, and that stone would just disappear. The other said the mountains appeared to be a frothy rock-like molten rock that has foamed up and solidified and would just crumble beneath you the minute you touch it.

Well, how are we going to design a landing gear for a place where half of it is dust and the other half is nothing but spongy rock? We pointed out that what we did know of the Moon from photographs showed clearly that there were some things like big boulders sitting on *something*. We said, "What's it sitting on?" They said there's a pinnacle under this great pool of dust and it's sitting on top of this pinnacle, this hidden mound under this pool.

What do you with that? These are Nobel laureates. We're just dumb engineers sitting there designing a landing gear. Well, we said, "They don't know what they're talking about. We're gonna go ahead just like it was Arizona." *[laughter]*

The people who were getting ready to launch Surveyor came by, and they understood the dilemma. They said, "What can we do to help you resolve this problem?" Well, that particular day I was in a very bad mood about something, and I said, "You can fly Surveyor right into the landing site, and if it smashes all to smithereens, we'll be home free." It was a terrible thing to say. These guys had come in good faith and offered to do whatever they could to help and I gave them this snotty response.

Well, they flew Surveyor and it landed and did not smash itself to smithereens. It took some wonderful pictures and later on, as you'll remember, the second or third Apollo flight landed right nearby and it was still sitting there. So, Surveyor ended the discussion about the Moon's surface.

ROBBIE: If Surveyor had never flown, you would have still gone on as if were like Arizona?

CALDWELL: I guess we would have. I don't think anybody had the guts to stop us. *[laughter]* Who was going to stop this whole great program just because somebody had said it was a pool, when any fool can look up there and know that anything that's been there that long and is floating around cannot be nothing but a great pool of dust. The world thinks that people who design spacecraft are great geniuses and know all about everything.

We decided that the lunar surface was the way we wanted it to be—the way that was convenient for a landing gear. I can't believe now though, that somebody somewhere in the system didn't blow some big whistle and call a great conference and do something to resolve this. These guys were not just guys off the street. You can't just ignore these guys, but we just kind of pretended they weren't even there.

ROBBIE: How long were you engaged in the design process for Apollo?

CALDWELL: Well, we had to get together with the launch vehicle people and the people that did orbital mechanics, and we increased the size of the team because it was getting very serious. Then we wrote a spec for Apollo. That was after the President had said we were going to do it.

Robbie: Who else was in the "we"?

Caldwell: Bob Piland wrote the specifications. His brother was the Procurement Officer for the Space Task Group and had come from a long line of lawyers and spec writers. I think the two brothers almost had a parting of the ways over that. Joe could not imagine assigning an important document like this to an aeronautical engineer who had never written a specification in his life. At this stage of the game, all the boilerplate of great government contracts wasn't the purpose. What you want to do is let the contractor know what you want. Later on, the lawyers can put in all the contractual terms. In the meantime, you want to get across the ideas of what you want, so you only do what you need to let the guy know the technical framework.

Bob and his brother came apart on that. I, personally, think that Piland was correct at that stage of the game. He went ahead with it, and the guys in my outfit contributed things. We went around and rounded up the life support systems guys and the propulsion guys, and everybody wrote their parts. Piland and I would edit them and put them all together. I don't think we had more than a ten-page spec for Apollo—ten pages!

Ken: When you say "Apollo," are you talking about the command service module or all of Apollo?

Caldwell: The command service module and enough description of everything else to put it in context so the whole thing would play together. Marshall was already designing the Saturn. The launch facility people at the Cape were gearing up what they would have to do. Nobody had a detailed design, but they were scoping the thing and figuring out how much money and how many people it would take.

Apollo was not yet a NASA program, but everyone knew it soon would be. Von Braun invited the Space Task Group to visit and discuss spacecraft/launch vehicle interfaces. Dr. Gilruth, Bob Piland, and several others and I were scheduled to visit. On the appointed day, we arrived a bit early, but without Dr. Gilruth—he had been called to headquarters at the last minute. We were directed to the large conference room at MSFC. No one else was there, so we sat near the center of one side of the great table, in no particular order. Bob Piland headed our delegation.

Soon the room began to fill with people. They didn't know us, and we didn't know them. They seemed unsure of where they were to sit. Someone of obvious rank entered, took the chair at the head of the table, and announced that Dr. von Braun was unable to attend the meeting. His presence and announcement seemed to clarify the seating arrangements. After a bit of shuffling, all the MSFC people took chairs on the right hand side of the table and the wall behind.

Greetings and introductions were getting underway when Dr. Von Braun entered the room and announced that he found he could attend the meeting after all. As one man, the chairman and every person seated to his right stood up, literally clicked their heels, moved one place to their right and sat down, leaving the head of the table to von Braun. The poor fellow at the other end of the table had no place to go! I like to think that Dr. Gilruth would have been proud of us for maintaining straight faces.

Putting Apollo Together

CALDWELL: There was a lot of work done at Langley on Apollo before we ever got to Houston. We let two or three "study contracts." One was to General Electric and one to Convair. There was a third to another contractor—I don't remember who. They were given the general requirements to get to the Moon and back, and not much else. Of course, at that time, the business of lunar orbit rendezvous had not been settled. In fact, we purposely avoided trying to direct the contractors along any lines. In the meantime, we were doing our own studies in-house.

Radiation hazard was a big fear at that time. People were very worried about the Van Allen belt, and it was thought that you needed some sort of protection from radiation. General Electric went overboard on that. They had what they called a "safe haven" in the spacecraft. It was a little storm cellar, you might say. I presume it had lead shielding and all this sort of thing. Of course it was heavy and it took a lot of space, but they were reacting to this fear.

Everyone had a different type of re-entry module with a different shape. There were three or four favorite approaches. One was developed at Ames—a "lifting body," it was called. It looked like a blunt-nosed half cone. It would have performed very well aerodynamically but it was difficult to package a crew in the thing.

There were various kinds of schemes for landing on the Moon and getting off. The study contractors avoided making any kind of decision about exactly what to do when we got to the Moon, other than in a vague way that something went down and something came back up. They may have not wanted to commit themselves to something that could have turned out to be wrong, so they kind of wishy-washed around it. Anyway, I think we probably stole a few ideas from someone, but by and large, we decided to stay with the thing that we had all worked on.

It was somewhat unfair of us, in a way, that we had the advantage of knowing what the three contractors were doing but they didn't know what we were doing. Also, we were the boss, so it really wasn't a fair fight, but I think it turned out for the best because some of those contraptions, I'm pretty sure, wouldn't have made it to the Moon and back.

KEN: On the discussion of the re-entry vehicle and the various shapes, was there a strong influence from the previous unmanned ballistic missiles? Was that a factor at all in your design?

CALDWELL: No, it wasn't, because the accelerations were very high with the ballistic missiles. One of the reasons for this blunt body is that you could manage to decelerate over a longer period of time and the heating was not as severe. The benefit of the lifting body was that it was very easy on the crew because they could essentially glide in, as opposed to ramming it. But it had that disadvantage of being asymmetrical. All the things except the thing we were promoting were pretty much asymmetrical. There was nothing so bad with that, but we had been in the pilotless aircraft business a long time and had had ever so much trouble with things that were asymmetrical.

KEN: Then your previous flight test experiences had pointed up those problems.

CALDWELL: Yes. We probably could have gotten along just as well without the studies, but it served a political purpose, if nothing else. It gave at least the appearance that we were not just ramrodding things through, though I suspect we were.

ROBBIE: But you said you *weren't* ramrodding it through.

CALDWELL: Well, if it looks like a duck, walks like a duck, quacks like a duck, it's a duck!

We finally got the proposal out. We talked before of the difficulties of writing the Request for Proposal, and the problem that Bob Piland had with his brother on the legal wording. Bob would say, "We can get to that later. The important thing, now, is to get the technical concept across, and then we'll worry about what's legal and what's not and about contractual obligations and all that." But his brother, having worked on the other side of the house, was more interested in making sure that the contractual end was legal. Anyway, we put it out without him and I don't think it was more than twenty pages.

ROBBIE: Would you refresh our memory about exactly how you all arrived at this blunt body design?

CALDWELL: We had found that the blunt body was a very efficient way of slowing down a spacecraft as it entered the atmosphere without developing too much problem with the heat shield. The reason is that it created an enormous amount of heat, but the heat was separated by stagnant air, so that the hot air got shed because the blunt body keeps, in a sense, a little pad of stagnant air in front of itself all the time.

Ken: Like a cushion.

Caldwell: There is new hot air that comes along. This thing just keeps one real blanket of air that tends to cool itself down so the spacecraft itself does not feel all the heat.

Robbie: I think you said that if you made missiles with pointed noses, the point would melt and become blunt anyway.

Caldwell: That's right. It would become blunt, and after it achieved a certain amount of bluntness, then it would create an equilibrium situation. But all of that led to things like the launch escape system. Some of the other folks' studies didn't deem that very important, but we had gone through Mercury with it and it ended up with the launch escape system. So that's another reason. When putting a launch escape system in a command module, that's a body of revolution that's symmetrical, you've got a much more pleasing aerodynamic situation than if you try to put it on something that, let's say, looked like the Shuttle. Now, Alan Kehlet had his favorite thing, which was called a "lenticular shape." It's like a lens or like a discus. He promoted the discus shape because it could serve two ways—it could be a blunt body when it came in flat or it could be a lifting body if it were tilted over a little. But it had the terrible disadvantage of not being very thick.

Ken: Without enough room for the crew.

Caldwell: All things considered, a ball or a sphere is probably the ideal thing. You get the most space in a sphere for the least amount of space outside.

Robbie: Why couldn't you do the sphere?

Caldwell: Well, for one thing, the sphere will not develop any kind of lift at all and there are guidance problems in returning to the earth and hitting the right spot. Without pinpoint guidance, from the Moon you would have to generate some sort of lift force. A sphere is neutral in every direction. That was a big argument for this lifting body. The lifting body was a fairly efficient flying machine.

Robbie: And you could direct it better. You could point it where you wanted it to go?

Caldwell: Well, the others could have been pointed all right. Some of the guys who promoted the other approaches said, "You can't develop *enough* lift. You're dangerously close to the edge of how much lift you need." It turned out later on, though, that the navigation and guidance was much better than people thought, and, luckily, it was, because the old command module, by the time it was flyable, would not develop as much lift as it was intended to have.

One time we even considered gold ballast, because it was heavy. There were several other metals as well, but gold is pretty damn heavy. We probably spent more money trying to figure out other ways of doing it than it would have cost to use the gold. But if it had been a lot of gold ballast, wouldn't it have been a lot of fun to see all the pirates that tried to steal it?

Robbie: That would have lent an air of swashbuckling romance to the space program!

Caldwell: It sure would have. It would have made darn sure that *someone* found that capsule!

Robbie: You know that article that came out in *The New Yorker* several years back that talked about you and Faget? The way the author portrayed it was as if you and Faget worked in total symbiosis with each other and it was just the two of you doing all this stuff, and I just wondered if the article is accurate. Did he overstress the relationship between you and Max or was it as he portrayed it in the article? *[Annals of Space: Max Faget and Caldwell Johnson,* by Henry S.F. Cooper, *The New Yorker*, Sept. 2, 1991, pp. 1–69.]

CALDWELL: He probably oversimplified it because, over time, our relationship would get closer and then spread apart and then get close and spread apart, just from an organizational standpoint.

KEN: It was dynamic.

CALDWELL: Yeah. Sometimes these things depend on your relationship within the organizational structure. If you're in different divisions, your relationship is different than if you're in the same one. Sometimes the assignment you have is identical and is something that brings you together. Sometimes it isn't. For instance, in the writing and putting together of the Apollo specs, I was reporting to Bob Piland, not Max. It was an entirely different outfit.

ROBBIE: At what points did you work most closely with Max?

CALDWELL: In PARD and in the early days of Mercury, and then it gradually drifted apart and then it came back together again. Now when I say we were apart, I don't suppose we ever worked more than fifty feet apart. It was one small family anyway. It's not like there were hundreds and hundreds of people there. So I don't suppose that many days would go by that there wouldn't be some conversations, one way or the other, between Max and me.

"North American's Got It Wired"

CALDWELL: Alan Shepard was a member of a panel that I chaired, one of several panels charged to evaluate the technical merit of the initial Apollo contract proposals. Try as I might, I could not get him to sit down with the rest of us and jointly evaluate each proposal on the basis of criteria furnished to each panel. He had other things to do, but I said, "Come on, Alan, it's tough work to read all these things and then we sit down and we discuss them, but we've got to evaluate these things. It's not fair to these guys who put these things together that we don't even evaluate them." He said, "I know, I know, I know. But North American's gonna get it. There's no use even bothering with it." He said, "North American's got it wired."

I said, "You don't know that and I don't know that, and even if they have, we've got to do our duty." He said, "Nonsense! They've got it wired." He'd occasionally show up, but after about ten minutes, he'd get a cup of coffee and you'd never see him the rest of the day. Anyway, I said, "Alan, why are you so sure?" He said, "The West Coast needs the money. That's one reason. They just got through with the X-15 and they're flying high, wide, and handsome right now, and they've got a lock on it."

Well, sure enough, North American got the job. Incidentally, I don't know what all the scores totaled, but they didn't come out so red hot on ours. He was right, though; we *were* wasting our time. I have no quarrel with North American, though. I think they did as good a job as anybody would have done.

KEN: Was it later that the lunar module was bid or was that part of this?

CALDWELL: Yeah. It was just about that time that we moved to Houston. What a mess! Here we were, through with Mercury, Gemini was getting underway, they were starting bids on Apollo, and altogether there must not have been a hundred of us. We were moving to Houston. We didn't have a Center to go to. They had rented office buildings scattered hither and yon around Houston. We were trying to staff at the same time and all this mess was going on. It was chaos!

On top of that, Robert Seamans was the administrator. He was an old MIT grad. Bob Chilton was head of Guidance and Control in our group, and he was also an old MIT grad. Charles Draper, at the time, was also riding high, wide, and handsome.

The first thing you know, somebody was convinced that the guidance and control contract should not be part of the Apollo program. North American, the general contractor, had won the entire contract, all except the guidance and control. They had the overall responsibility, except for this one vital part.

ROBBIE: Whose idea was that, to separate them out like that?

CALDWELL: I guess it was the MIT people. As far as I was concerned, that was the least of my worries—the contractual structure of the thing—but it did, later on, turn out to be a troublesome situation because North American could not control one of the vital systems. They could not force anybody to do anything to integrate it with other things. It wasn't all that bad, because Bob Chilton was a sensible person and other people were sensible and there never really was, as far as I know, any big problem, but it was a most odd arrangement.

ROBBIE: It sounds like their first dysfunctional decision.

CALDWELL: Yeah, but it worked, I guess, because the people just simply made it work. I'm sure North American must have gnashed its teeth many times over this. Then of course, before long, all of a sudden here comes Grumman with the lunar module, completely independent of the other two. So you have to set up a new Program Office to deal with that. But somehow it all worked out.

Anyway, we had no office that the contractors could start with. Not only did we have no offices here in Houston, we had no staff here, no on-site representative. Of all things, my secretary set that up. I sent her out there and she said, "I don't know anything about it." I said, "Well, you know more than I do about it." Thankfully, she got involved with some other young lady out there, a secretary at North American, who knew how these things went, and between the two of them, North

American donated office space and typewriters and telephones and that sort of thing and got that set up. One small incident shows you how things worked. Continental Airlines knew that something was going on there because the people going to California from Houston jumped about tenfold in three days. The head passenger guy called me because I was the so-called Head of that Apollo-North American contract and said, "What can Continental do to make life easy for the many people that would be travelling back and forth?"

Well, like a fool, I presumed to answer. I should have known better than to get into something I had absolutely no knowledge about, but I said, "We get off at 4:30. Just give guys the time to get to the airport and get an airplane and get us into Los Angeles at a reasonable time at night." He said, "Great!" And I said, "Our per diem goes in quarters of a day, so 6 o'clock, 12 o'clock; that's how we divide it."

So he arranged the flight at about 6:30. It was ideal. A few weeks later, he called and said, "What in the world is going on? We don't have anybody flying. They're all flying United." I said, "What happened?" He said, "United supposedly leaves at five minutes to 6:00. All your guys are going on United because they get another quarter day per diem if they leave then." That taught me a lesson: "Don't mess with something that you don't know anything about!"

Once, Charlie Frick and I were in his rental car, trying to make the last plane at night from LAX to Houston. It appeared hopeless. Charlie pulled up in front of Continental's departing passenger entrance, switched off the ignition, but left the key in the lock. He said, "Let's go! If we make it, good. If we don't, we'll have a car waiting." We made it. Some weeks later our travel office asked me if I knew anything about an *enormous* rental car bill charged to Mr. Frick!

Another time, a North American helicopter flew Shepard, Slayton, and me from Downey to LAX for a midday flight to Houston. The agent at the check-in counter recognized Alan and Deke and changed their coach flights to first class, then looked up at me. The other two said I was with them, and he changed my ticket to first class too!

The only people in first class were two hostesses, the three of us and a "Little Old Lady from Pasadena" [a song recorded in the 1960s by Jan and Dean]. Our two hostesses, with the help of the two assigned to serve the dozens of passengers in coach, showered us with attention, food, and drink. The little old lady finally got the attention of a hostess long enough to complain of being cold. The hostess flung her a blanket and turned back to chatting with Deke. The entire trip went like that, and as we were deplaning in Houston, I asked Alan if it was always like that. He said, "No, sometimes better!"

Anyway, it was kind of chaotic in those days. That situation created a lot of domestic problems. There were a bunch of guys that had come from the tiny little town of Hampton where there were no bars, no nothing, whiskey was illegal, no drinking except in a private home, certainly no open bars. Here we get to Houston, and half of the wives were still in Virginia. Houston was sort of a wide open, freewheeling place. We were traveling all the time on airplanes back and forth. I think it strained several marriages to the breaking point. But they settled down after a while. Everybody got their fill of just carousing around. But it was rough going. People worked hard, but they played hard too and a lot of times played with people other than their wives!

There was an Apollo Program Office, and there was the Engineering Directorate, which Max ran. Charlie Frick had come from Convair and ran the Apollo Program Office here in Houston. Piland was his deputy. I was in charge of the command service module. Technical support was vested with the Engineering and Development Directorate. The idea was that if a technical problem arose, the Program Office would seek one of Max's experts in a particular area, and that guy would furnish the expert piece of technical guidance that you needed. The authority to issue changes came from the Apollo Program Office. I didn't think much of that from the start, but my point of view was biased since I was in the Program Office. I knew immediately it would be a short circuit between Engineering Directorate and the contractor—a short circuit through the Program Office—and the Program Office was ultimately responsible for the money spent and the schedules. It was set up and before you know it, the subsystems manager and his guy at the contractor were forming a liaison that was stronger than within the government itself, and next thing you know, the Program Office was dealing both with the guy next door to you and the contractor. Now don't get me wrong; it worked! But it was a trial, and often it would depend largely on personalities and who was strong and who was weak to make it.

The subsystems people were cooped up in the Rice Hotel to hammer out their own particular specifications and their budgets. They were literally locked up in these rooms and told, "You work this out."

The two sides had directions from their management that were not the same. [*laughter*] North American, the contractors, of course wanted to get everything out of this they possibly could. All of us were naive. Half of these guys doing this had no idea what things cost. They'd always worked for the government. The contractor's guys were told to get every penny they could get, no matter what. You really can't negotiate under such circumstances. There's no place to negotiate; you have no sense of values.

But they were forced to do it and they did it, and some of them came out of this as mortal enemies. After being cooped up for a month and being forced to try to negotiate and not being able to, they came out despising one another and it lasted through the program. Some of them came out of there buddies, which was almost as bad. It was just a terrible way to do things. The contracts should have been negotiated by professional contracts people, not engineers. Engineers have no idea about money and cost. Not only did it create this ill feeling, but for at least a month, it took the very people that you needed at those times to get the thing going.

Before long the BellComs and the General Electrics and these various groups of people said, "What the hell's going on. Fifty billion dollars and we're sitting up here not getting a piece of this!" They all would come in and say, "We can help you do this. We can help you with that."

All of a sudden they woke up to this and Charlie Frick was beating them off with sticks as best he could. Of course they were calling their congressmen about this "son of a bitch at Houston" who wouldn't even let them in the door, and Charlie said, "I'm not long for this world, but if you guys can just get this thing technically squared away before they get hold of it, that's all I want." Sure enough, that's exactly what happened.

I've got a line down here on my notes that says, "Carpetbaggers." That's what we used to call these people.

Fifteen-Plus Pure Oxygen

CALDWELL: Let me say something about some technical issues. Probably the most important thing of all—and this has to do with the great fire—was cabin atmosphere. The cabin atmosphere was settled at Langley to be five psi [pounds per square inch] pure oxygen. Everyone recognized that that was a dangerous situation. In fact, some tests had been made in a pure oxygen atmosphere, and several guys had been burned to death. The problem was that the pressure garments at the time, at anything greater than 5 psi, were practically immobile. A man dressed in them could hardly move. The garments mostly were just an outgrowth of aircraft pressure garments in which it was not necessary that the pilot do things like walk on the Moon or climb up and down ladders. He had to flex his arms to some extent to fly the airplane, but it didn't require a lot of mobility. It could only take five.

For some reason, the medical people said the crew needed five psi of oxygen. On Earth, the pressure of oxygen in this room is five psi, out of the total pressure of 15. Nitrogen makes up most of the rest of the pressure. People in Mexico City and Denver only get three psi of oxygen and they get along fine. Now there's no doubt that if you suddenly go from here to Denver, you'll feel a shortage of breath for a while until you get acclimated, but the medical people said the guys needed five.

We were worried about leakage in the spacecraft for the 14 days they would be there and being able to carry enough make-up gas if we had a leak. Of course, the higher the pressure, the more it's going to leak. The higher the pressure, the stronger the command module has to be. Everything considered, the decision was made to go with the five-psi oxygen. None of us, anywhere, ever seemed to think about the fact that when we tested this thing on the ground before launch that it would turn into 15 psi.

ROBBIE: What do you mean?

CALDWELL: The 5 would go to 15 because we're sitting in an atmosphere of 15 pounds per square inch here and to make it all oxygen it has to go on up to 15 or there would be a negative pressure and you would collapse the structure. There is no atmosphere in space, so now we've got 15. In fact, just a little more than 15, because you go above 15 to see if it's leaking.

KEN: And we're talking about testing the real article. It wasn't like it was a prototype.

CALDWELL: And with guys in it before launch. In thinking about this later, I cannot believe that none of us thought of that situation—I think we were so focused on the flight situation. Fifteen-plus pure oxygen is one hell of a lot different than five. What's more, sitting here on the ground, nitrogen, which is inert, is all around. Most of the atmosphere is an inert gas, so that if the oxygen consumes this piece of paper and burns it up, it's used up the oxygen and the rest is nitrogen, so nothing happens. If you have got the whole room full of oxygen, everything is gonna burn. *Everything.* I can't believe we didn't think of it. North American, from the beginning, protested the pure oxygen atmosphere. They said we should put in a system that would dilute it.

ROBBIE: Dilute it with nitrogen?

CALDWELL: Or helium.

ROBBIE: Would that have made it less flammable?

CALDWELL: It sure as hell would have. North American was aware of this thing. They protested, and on several occasions we denied their protests. They finally sent a formal letter to Bob Gilruth, in which they said they wanted to go on record as protesting this—that they thought it was dangerous and they would do it if directed, but only if *specifically* directed to do it. I've never heard anybody in NASA mention that, but I know it happened because I got called in with several others for a meeting in Gilruth's office to discuss this thing, and the decision was to stay with the oxygen atmosphere.

ROBBIE: Why?

CALDWELL: Because at the time, the technology was not up to doing anything else. It was, but we didn't know exactly what to do at the moment to have it do it. It could have been done and, in fact, it was done later on. But the decision was made. I personally sent the telegram to Rockwell that directed them to continue with the pure oxygen atmosphere.

ROBBIE: You must have felt awful.

CALDWELL: I did! It was done in good faith, but it was a *stupid* error. I have always felt badly because Harrison Storms took the lumps, and North American took the lumps. Now, it's true that their sparks with crummy wiring ignited the system, but it was a bomb waiting to go off. It's unfair to say it was their entire fault because of the spark. NASA either didn't know it or chose not to know it, yet there were people there that knew it. I certainly didn't make the decision on my own to do it. Somewhere in the deep files there must be a telegram or something.

ROBBIE: From North American noting their protest.

CALDWELL: And my reply to it. It was a terrible, terrible, *terrible* mistake and there's no way to undo it. The point I'm making about this story is that a handful of people made the decision. We did not convene any great body of people and send out announcements.

Robbie: So the handful of people was Gilruth and you and who?

Caldwell: Must have been Max.

Ken: Probably Bob Piland.

Caldwell: Piland. You know, it certainly wouldn't have been just Gilruth and me.

Robbie: And all of you decided to go with the oxygen because you knew how to do that, but you didn't see the possibility, at the moment, of developing the technology to put the nitrogen into it.

Caldwell: Right.

Robbie: And did you talk about the hazards of fire? What was the conversation?

Caldwell: Well, I can't recall the details. We had gone over this any number of times before North American protested.

Robbie: And you recognized the flammability.

Caldwell: Oh yes! As I told you, we were aware of that accident in a similar situation at Brooks Air Force Base.

Robbie: Did you all just decide, "It's just not going to happen?"

Caldwell: Yep.

Robbie: Then, knowing that risk, why weren't people more careful with the wires?

Caldwell: What I'm really trying to say is that we never envisaged that sitting on the launch pad, the cabin would be pressurized to 15.

KEN: So you were so concentrated on the flight that you forgot what happens on the ground.

CALDWELL: And the people on the ground who were in charge of it assumed that the people doing this knew what they were doing. So it was just a case of a terrible mistake.

ROBBIE: With North American's letter, were they complaining about the 15 on the ground?

CALDWELL: Yes.

ROBBIE: So they *did* make you aware of it.

CALDWELL: Hmmm. Wait a minute—I'm not sure of that.

ROBBIE: So they may have simply been protesting the pure oxygen.

CALDWELL: I couldn't say—I don't recall. I don't think I, personally, read the letter, but I know they formally protested.

ROBBIE: They formally protested pure oxygen.

CALDWELL: Yes.

ROBBIE: But it could have been that everybody was still just thinking about space.

CALDWELL: Could have been. I don't know what North American was thinking.

ROBBIE: So you think North American didn't publicize their warning letter because...

CALDWELL: No, no, no, I'm not going to say that at all. I was just speculating that because Ken had wondered why North American didn't say something about their protest—I have no idea why they didn't do it. I'm sure they must have had some reason for it. Well, it was a terrible thing.

Let me give you another case, which is landing impact. We all knew that it was quite possible that the command module would come down on land instead of water. We had to provide some sort of impact attenuation. We couldn't just let this thing hit the ground—kaboom! It's bad enough hitting the water, but hitting the ground is worse. North American wanted to drop the heat shield on hydraulic struts, but I had had enough trouble with the dropping of the heat shield on Mercury and that airbag, seeing all the trouble that that thing caused, that I adamantly opposed doing that. I thought what they should do is spring the individual couches for the crew.

ROBBIE: Spring the couches?

CALDWELL: In principle, I argued that attenuating the impact with a 200-pound man in the seat was a lot easier than a whole 14,000-pound command module. We didn't give a damn whether we busted up the command module—the hell with that—we just wanted to save the men. But North American had already done a preliminary design. Charlie Feltz was their chief engineer. He and several of his guys and I argued the issue all day long. It got up to suppertime and we were still there arguing. All of a sudden Charlie said, "I've changed my mind. We'll do it. We'll attenuate the seats." That was that! Now, the reason I'm telling you this is that there was no "Change Board," there was no great meeting. There was Charlie Feltz and a couple of his guys and me, and we sat there and we argued the thing all day, and when we got through they said, "We'll change it" and they did.

ROBBIE: That's a functional system.

Caldwell: Yeah. We could have screwed up but we didn't, so I don't guess we'll ever know. It never landed on land, so we don't know. And that wasn't a serious problem, springing those seats. It's a small space in the air and you have to have room for it move up and down and sideways and this kind of thing. It had its troubles, but it would have been one helluva mess to spring that whole damn 14,000-pound spacecraft.

New Territory: Space Crew Quarters

Caldwell: Somewhere along the line, I inherited the old Advanced Spacecraft Technology Division, which I was trying all the while to reconfigure anyway. I wanted to work on more current things, and we were starting to talk about the Shuttle, and I didn't like the idea of designing space stations that were never going to be built.

Ken: These were the early space stations.

Caldwell: This was *way* early. The present one bears no relationship whatever to these things. It was one of the divisions under Max. I talked Max into reconfiguring the thing and we changed the name to "Spacecraft Design Division." Now I felt more like I was at home.

About that time, Skylab was getting under way. Skylab was taking pieces left over from the Apollo program to essentially be a small space station of its own. The responsibility was assigned overall to Marshall Space Center. JSC, because of the crew, had things to do with crew training and some medical things to do with the crew and things like food and clothing and that kind of thing. JSC paid absolutely no attention to it.

There was a guy called the Principal Investigator for Habitability and Crew Quarters. His job was not to *do* anything but just to *report* on those things that would affect crew performance and living conditions on the spacecraft. He became ill. So I was told to take his place, to be the Principal Investigator for Habitability and Crew Quarters. All of a sudden, I got very interested in this.

KEN: This was new territory. It hadn't been done before.

CALDWELL: That's right. We hadn't worried about crew quarters. Hell, it was just three guys sitting in a damn little place and sleeping in the same place and defecating in a glove in their own hand and eating out of squeeze bags. Now, we've got a room and all kinds of possibility. But I didn't really have any responsibility to do anything about it. All I was supposed to do was look into it and later on report on how it worked. Well, I couldn't take that. I went to Marshall and I found out, for instance, they had sleeping bunks on the wall, hanging with chains, like on a submarine or something. In zero gravity!

ROBBIE: They'd just float around, wouldn't they?

CALDWELL: Well, of course! They had no provisions to eat at any one place. They didn't have any toilets. They didn't have anything! They'd set aside a room. It was clear that the guys there were uncomfortable with this, that they had no organizational structure that dealt with things like this. Their business was jet propulsion systems.

ROBBIE: So why were the guys at Marshall even working on this?

CALDWELL: Well, because they got assigned to do it. It was put in there and they were damned if they were just gonna voluntarily give up something. I went down there one day and a guy was painting over the lettering on the door that said "Propulsion" and was writing "Crew Systems" on it. I'm not kidding you!

I came back and I must have raised hell at the right places. This was ridiculous! Here we were going to put up three sets of three guys for God knows how many days and they were going to have to live under these conditions? I convinced everybody that something had to be done, so all of a sudden from my role as investigator, I became the designer, which is what I like to do. It wasn't quite as easy as you'd think, though. Marshall didn't quite take this lying down, except that they were really kind of relieved. I had developed a document that I called "Initial Guidelines" for this thing. I said, "I'm not gonna just report it. I don't want to report that something *doesn't* work. I want to report on something that *does* work." I said, "What is habitability?" Well you know what it is and I know what it is, but just try to *tell* me what it is. I decided to break it down into elements so I could deal with it one element at a time, and I picked the respirable atmosphere—the air you've got to breathe. I put that first because of that terrible thing in Apollo. I may have unconsciously put it there. The architecture, the physical arrangement of the room. Clothing, mobility, and restraint, food, waste management, recreation. I listed these elements and there were nine. That way you could take each of them and try to say what you were going to do to make it better.

Nobody protested, so we started doing it. Things like the garments, for instance. Before, everyone had worn flight jump suits. There could not be anything worse! It's the worst possible thing you can wear.

Robbie: Why? Because if you had to go to the bathroom you had to take everything off?

Caldwell: Yeah. Besides, from the crotch to the shoulders is a fixed length. We've got nine men who are going to fly. We don't know who they are. We'd better put the provisions on board before they even get there. There are the variations in size, for one thing. How would you like to be 6 foot 6 and have a jumpsuit that's for a guy who is 5 feet 6? It would not be a very good thing. I said, "Besides, the temperature situation will fluctuate and it's ever so much easier to take your shirt off or put on shorts or something like that. You can adjust yourself. We've spent thousands of years on Earth learning how to do this. They said, "Oh, that's what we use all the time!" Well, these guys were flying airplanes all the time. There can't be anything worse than going around in zero gravity with big, cumbersome things on your legs. You're going to inadvertently bump things all the time with your legs anyway. You want pockets just exactly like you've got in your clothes now, except maybe we put a bit of spring stuff in there so that when you push your hand and pull it out it closes, so things don't come out on you. Get these same clothes—buy 'em at JC Penny.

ROBBIE: Did you spend time in zero G to figure this out or did you just know it?

CALDWELL: I just knew it. I got all cleared to do that in the KC-135 but never did get around to doing it. Finally, I talked everybody into doing it. The crew accepted the regular pants and shirts and socks rather gracefully, but they finally insisted on things with their patches and their ranks and their serial numbers.

ROBBIE: They wanted to look like astronauts?

CALDWELL: That's right, and the clothing we had designed didn't have any of these things on there. In fact, it was NASA blue and had a white stripe along the legs and arms so that when you photographed things it was easier to see the articulation mechanics of the body. It looked fairly attractive, but the crew just absolutely rebelled at those white stripes and that blue and insisted that it be colored khaki. But that was all right. We'd at least won the battle of getting pants and shirts.

One of the shirts, an undershirt, was made out of some material to resist fire. Everybody was always worried about fire, so the choice of fabrics was limited. They would pick things that were not so flammable, and that was too bad because it's hard to beat cotton for comfort. But I think there was an undershirt made out of polyester of some sort, and the crew would constantly complain of how everybody would stink after wearing those undershirts. Now, I've noticed that on the shuttle there's no problem with the cottons. Of course, the shuttle has a regular atmosphere, so the fire hazard is less.

As for eating, we all agreed that there's no way you're going to spend three months squeezing stuff out of a bag to eat. Sooner or later, you're gonna stop eating things because it'll just get so distasteful. With our technology, we know how to deal with putting stuff up in cans just fine. We can have freeze-dried foods. We designed that little tray that you can put these small cans in, and have a timer and an electric switch on it so that you can put it in there in the morning and by the time you come back, you know that people are wearing more or less what they're used to wearing themselves—for lunch let's say—they've all been warmed up.

We used knives, forks and spoons, and everybody said, "Zero G? Knives, forks, and spoons?" Why not? Oneida gave us some knives, forks, and spoons that were stainless and magnetic, and we had magnets in the little tray and it worked just fine. They've been using it to this day, and as far as the zero G, half of the food you eat, whether you intend it or not, has got gravy or something equivalent to gravy with it. You can take your plate and turn it up like this and most of the time stuff won't even slide off in one G. Now in zero G it's certainly not gonna fall off. I wouldn't advise trying to eat green peas, though. They'd fly away. Remember the old thing, "I eat my peas with honey. I've done it all my life. To you it may seem funny, but it keeps them on my knife." That's what got us started on that whole business with the gravy—that poem. Oh, and these little trays were made to hook on a little pedestal so you could make a table, so then three guys could sit around this table and eat, instead of hooking themselves somewhere and hanging on to something and squeezing bags.

As far as the architecture was concerned, there needed to be sleep stations and a place to work and a place to eat. It had just been one big room, so I devised a thing up with some compartments and I said, "We've gotta have a thing we'll call the 'ward room' where the guys eat. Not where they sleep, where they eat."

The people at Marshall said, "Oh my God!" They were over-budget and over-weight and behind time and here I was, coming along and putting in some more. Then, when they got all through I said, "We gotta have a window in the ward room. When you're sitting here eating, you've gotta be able to look out through the window."

Robbie: So you were really thinking about quality of life.

Caldwell: Yeah! You're right! These guys have to stay there three months. They've got to have these things and you've got to reduce the daily living to something incidental to the job. You don't want to have to focus all your time on surviving or you won't have any time left to do any work.

Suppose you had to spend your whole day figuring out how you're going to sleep, how you're going to eat, and how you're going to wash. We made some little things that would dispense water and suck it back up and rags you could squeeze so you could wash. We found out that you can shave and put lather on your face exactly like you do on Earth. The lather holds all the shavings together and you can wipe them off with a tissue.

Robbie: And did you find that out by having someone experiment at zero G, or what?

Caldwell: Well, we'd check them out, but you could just reason it out. However, there seemed to be a point made of not doing it the way it's done on Earth. Like the latches on the cabinets—

everybody was inventing a latch. I said, "Ace Hardware has been doing this a hundred years. Why in the world do you have to invent a latch?" "Well, it's because it's not space qualified." I said, "Well, qualify it. Buy some and then qualify them."

The window was really a low blow to Marshall, to come up leading the program and insist on a window. They just said, "No, you cannot do it!" NASA had hired a famous French industrial designer named Raymond Lowey. Marshall wouldn't have anything to do with him. Well, I found out about it and I said, "As long as he's gonna get paid as a consultant, I can't think of anything better than to have the advice of Raymond Lowey."

He was delighted! He just couldn't wait to do something on the space program. On the first visit he made to Houston, he arrived one morning when the fog was so thick you just couldn't see anything. I met him at the airport and we came through the fog and he never saw a thing until we got inside the building. During lunch, we went over to the cafeteria, where Gilruth had asked us to have lunch with him. We had a polite conversation and Gilruth said, "Mr. Lowey, I hope you got to see the center, and I hope you enjoyed it here." Lowey said something very polite and when we left the lunchroom, Lowey turned to me and said, "I didn't say anything, but if the outside looked as dismal as the inside, I'm just as glad it was foggy this morning." Well, George Mueller called a great meeting at Headquarters to resolve this business of the window. He appreciated the problem in terms of what it was costing Marshall, but he also understood that it would be a great thing to have in the spacecraft. So the folks from Marshall got up and made their pitch as to why the window was so bad and so costly, and I got up and just talked about how there were three crewmen with six eyes, and the eye is the most sophisticated sensing instrument that ever was invented. To deny the eye the ability to see what's going on outside is absolutely ridiculous.

Mueller, of course, was in a terrible spot. He was in a no-win situation. He turned to Lowey and said, "Mr. Lowey, what's your opinion?" Lowey said, "I can't imagine not having a window." Mueller just turned to the Program Manager, and said, "Put the window in." I'll guarantee you that I was *persona non grata* at Marshall after that!

ROBBIE: But I'll bet the astronauts were grateful!

CALDWELL: Oh yeah! They told me they never got tired of looking at the Earth, and fortunately, on Skylab, the window was mostly facing the Earth or the horizon, and several times they saw things that no one had yet explained—things like little sparks of light. Later, these observations were explained, but these phenomena would never have been noticed if they hadn't seen it from that vantage point. Seeing the Earth from that distance recalls the old expression, "You can't see the forest for the trees." Well, this is a case where you can see the forest because you're far enough away from it.

ROBBIE: Was the window in the wardroom the only window on the spacecraft?

CALDWELL: No, but all the rest were at a remote place and were designed for scientific purposes. They may have had cameras mounted in them, but they were not windows in the sense that you can just idly sit and look at whatever you please. They had some particular purpose, such as viewing the sun.

ROBBIE: So this was the only window that was just for the astronauts.

CALDWELL: Yes, and there's a good reason for that. Windows are terrible things to put in a spacecraft. They're fragile, unless you make them big and thick and heavy. They're thermally poor and transmit temperatures back and forth.

ROBBIE: Is there a concern for the possibility of leakage?

CALDWELL: Definitely. And to put one in after the program is way well along is just terrible. I knew that. I'm enough of a structures guy to know what a pain that was.

ROBBIE: I remember Max saying he didn't want to put a window in Mercury, but the astronauts rebelled and insisted, so he did.

CALDWELL: From a purely engineering standpoint, it's a dumb thing to do. I've gotten up a couple of times to see our present space station. They don't have any decent windows in their living quarters. Of course, everybody kept giving the same answer: "It's a pain in the ass and what are they gonna see?"

ROBBIE: It's interesting that you were an engineer and you knew from an engineering standpoint it was a "dumb" thing to do, but you were taking this humanitarian view about what works for the *people*.

Caldwell: I just put myself in their shoes. Besides, you see, up until then, no one had been in a spacecraft for more than 14 days, and the mission was so overwhelming that it was inconsequential whether a guy felt very comfortable for that length of time, so long as it didn't interfere with his duties.

Now the food was somewhat troublesome in that it was compromised a bit because, for medical reasons, certain foods were not allowed and certain foods were given to promote good bowel habits and certain ones were avoided for that reason. Maybe the calcium intake was supposed to be such and such. So everything didn't come out exactly the way you want it.

But the old waste management system worked better than the one on the shuttle. We never had any trouble with it. And the guys kept that place spotlessly clean. I was really worried about the accumulation of debris and waste here and there and dandruff and food particles smelling bad after several months, but it never did happen.

The living quarters were in the hydrogen tank of the S4B stage. Just behind that stage was the great oxygen tank. They were both empty by that time, so we put an airlock in the top of the oxygen tank, which was on the floor, you might say, with the airlock, so you could take your trash and garbage and open the airlock and put it in, close the airlock, open the other side and you had that whole big oxygen tank to fill with garbage. As a result, you never had to wonder what the hell to do with the garbage.

The Human Dimension: Saving Skylab

Caldwell: Then we had that terrible launch incident in which Skylab lost its solar arrays and its thermal shields.

Ken: That would be sometime in the mid-seventies.

Caldwell: The first 24 hours after it happened, everyone more or less was in the dumps and said, "Oh well, that's the way it is. It's tough but what can you do?" Then all of a sudden, kind of spontaneously, people in a lot of places said, "Why are we sitting on our hands? Let's *do* something."

We had an Apollo command module sitting there all ready to go, and we had a trained crew, and we had a great laboratory here, in a facility that had all the shops and equipment we needed. We had all the experts in the world right here, and just on a moment's notice, it got organized into an activity to *do* something. It's unbelievable how that center came together and Marshall came together after that accident. Everybody was anxious to help.

A guy named Jack Kinsler was head of the mechanical services division, and he and his gang had been working on different experiments with Skylab. He was aware that in the workshop part of the Skylab, there were some airlocks that one could put things into and push them on through to the other side. There was one right where the thermal insulation had gotten torn off during launch. Jack said we could make an umbrella type of thing, fold it up, push it up through this airlock and then open it, and it would make a sun shade where the insulation had been torn off. That was important because this was facing the sun and Skylab was getting hotter and hotter.

Some guys had analyzed the solar arrays. One was torn off altogether and the other was just jammed. It wouldn't open one way or the other. They decided that it was possible to cut what was holding it and remove whatever was impeding it, and then it would spring open by itself and start generating power again. Skylab had batteries but they wouldn't last for long.

So there was the possibility of getting the solar array to make electricity and to get the temperature back down. Well, both of these options were tricky things to do. How are you going to get a guy up there and cut these things loose? There were no tools made to do it. So, people started designing and built the tools overnight, and the crew started training to use the tools. Jack Kinsler and his crew got to work on the umbrella, building it in the shops at JSC. (Incidentally, he was one of the young model makers that I told you about who had come to work years and years before.)

Jack's scheme required that someone had to get inside Skylab first to activate the airlock from inside. We were afraid that it might be so hot in there that the crew couldn't go inside, so we said, "Let's have a backup." So I designed a system whereby the crew didn't go inside, but pulled an awning, hooked it on certain places and pulled it across. It was like a big sheet. Jack's scheme was a lot better and easier to do if it didn't get too hot inside, but we built mine as a backup. Two separate crews were working, then. Money was no object. Priority A1 AA, and whatever you wanted, you got.

One night we wanted some special paint to paint this awning we were making. The only place we could find it was in Durham, North Carolina. They got an astronaut to get in a T-38 and fly to Durham, pay the man his ten bucks for the gallon of paint, and fly it back to Houston. We painted the thing the next morning. I don't know what that gallon really cost, but we only had to pay the man $10 for it. And I'll be damned if it all didn't work!

Incidentally, this awning that I'd done required that a man open a hatch in the command module, fly up to Skylab, take a boat hook type of thing, attach the ropes, then pull the ropes to brace the awning up. Now, that was kind of dicey. We figured he could do it, but that was the second priority. Fortunately, when the crew opened the hatch on Skylab, it was only warmer than comfortable. They couldn't have stayed in there, but it wasn't so hot that they couldn't get in, and they rigged the parasol that Kinsler and his team had made, and it worked like a charm. The temperature immediately started going back down again. The crew went in and got out and cut some of the things that were impeding the solar arrays. The solar arrays opened and Skylab was on its way.

We were close to giving up, but somebody said, "We're *not* going to give up!" And we pulled it off. It was rather miraculous, and I don't think anybody ever paid that much attention to what a marvelous feat this was. It didn't counteract the terrible disaster of the Apollo fire, but it shows you that there are good things that can happen too. You can make mistakes but you can also bail yourself out of mistakes. I was disappointed that they didn't get to try my system, but I was very happy that the other thing worked.

Early Space Shuttle Concepts

CALDWELL: The shuttle was coming along about this time. That was what brought Max and me back together again. I was sort of out of work, or at least there was nothing real interesting going on. He was always designing things himself, and we started talking about a reusable spacecraft and how it ought to look like an airplane.

ROBBIE: Why did you think it should have wings?

CALDWELL: Because it needed to be maneuverable enough that you could land it on a regular landing field and not have to flop with the parachute in the water and all that kind of stuff. You can't reuse something that's landing in salt water. We were so ambitious at the time that we thought the launch vehicle should be reusable too.

ROBBIE: That would have saved a lot later on.

CALDWELL: That's right—it sure would have! He and I worked up a little design and we decided to build a small model of it that was light enough that we could demonstrate. Max used to be a model airplane builder himself in his younger days. He'd long since got out of it, but I was still building them. I'd always done that, so I had all the materials.

Max agreed that he was going to build the return vehicle and I would build the launch vehicle, and that both would have wings. The launch vehicle had two huge tanks, separated like a catamaran, and the return vehicle nested in between these tanks. We built the model and it was an aerodynamic disaster. Gilruth got wind of this and wanted in on it. He took one look and said, "You're like two schoolboys. You make the dumbest mistakes." He pointed out things about it that, aerodynamically, were absolutely silly. Things that we should have known but missed because we got carried away in our enthusiasm.

ROBBIE: Was somebody supposed to fly the booster on return?

CALDWELL: Oh yeah! Or maybe the booster could have been flown back unmanned, but it was supposed to fly back and land too. It turned out the spacecraft itself was not too bad, but the combination of that with the booster was just a disaster. All kinds of funny things were happening. I think Max still has the model. I intended to keep the booster part but something's happened to it, I don't know what. It was a balsa wood and tissue type of thing. That was the first shuttle!

KEN: Now, the point of all this is here we are today, getting close to the turn of the century and we're still trying to work this problem.

ROBBIE: Still trying to build a reusable launch system.

CALDWELL: That's right. The technology just wasn't up to it at the time. It wasn't too bad to bring the spacecraft back, but it would have been too tough to recover the launch vehicle.

ROBBIE: So what happened next? You were saying things happened almost the same way they did in Mercury.

CALDWELL: Yeah, there was no edict from Headquarters to start working on a shuttle program. It just got going, and more and more people became enthusiastic about the thing, and it looked more and more feasible—just about the exact same way that Mercury came about. Nobody said who was in charge of anything, but Max seemed to take charge of it. He was head of E and D and it was all E and D. No one else had any authority whatever to do anything on it.

The first thing you know, though, it got organized enough that Jim Chamberlin was assigned to put together a group of a dozen or so people, and they moved him into a loft in one of the buildings over there to set up a little task group to really get down to cases on this. As soon as that happened, there seemed to be a separation. Jim was not the kind of guy to cooperate with other people. His idea of an organization was a circle in the center with everybody on a radiant line that went out from it. Everybody reported to Jim directly.

ROBBIE: And what was he working on?

CALDWELL: He was appointed to head an on-site task group to develop more plans for the shuttle. I don't know if it was being called "Shuttle" yet, but a "reusable spacecraft." He had people drawn from different organizations—some from mine, some from others. Max and I were kind of pushed to the side and insulated. Jim made sure that we didn't get to talk to anybody but him, and he kept us at bay. Neither of us liked that idea, but I presumed Gilruth must have appointed Jim.

Don't get me wrong—Jim was a very capable engineer. He came from Avro. He was their chief aerodynamicist and he had run the Gemini Program. In fact, he was the first one to put order into the Mercury Program. He didn't like the haphazard way work was being done between McDonnell and the Space Task Group, and he set up a systematic way of doing things. Now, it turned out that this systematic way of doing things always put Jim in the place where he was the top of the system *(laughter)*, but Jim wasn't a bad ol' guy. In the meantime, Spacecraft Design Division was working, too. Max wasn't about to just turn this thing over altogether to Jim, so he had my division working on it too.

The Air Force got into the act as well. It was pretty clear, politically, that without the Air Force's support, money would not be available for this sort of thing. The Air Force had need, or at least thought they had need, for a ship like this too—a spacecraft. The whole thing had been designed with a very large cargo bay to start with, because it was reasoned that this so-called "shuttle" would be used to carry up elements of space stations later on. It was to be a ferry of sorts.

When the Air Force got involved, as part of their cooperation, they insisted that the cargo bay be enlarged to 15 feet across and 60 feet long. That was a sight larger than what we had anticipated, and changed the whole game. The Air Force also said they wanted to be able to come back to Earth and land at a given spot in one orbit from the word go.

Well, we couldn't come close to doing that. The performance required to land at any place on Earth from any place in orbit in one revolution is a helluva lot different from taking two days to do it. If you notice, the Shuttle doesn't just come back right away, it starts off and gets prepared and it takes more than one revolution to come in. Anyway, that forced a great change and it almost threw everything out the window that had been done before.

I think at one time we counted 41 configurations, as we kept changing and modifying, working all the time with Marshall who was doing the launch vehicle. It was Apollo all over again. They let some study contracts—I don't think anybody paid much attention to them. Probably it was a political gesture. People were very much upset with in-house activities. The commercial enterprises, understandably, did not like the government doing work in-house when it could have been paying *them.*

As a matter of fact, if you look at the Johnson Space Center, you can see from its beginning that it was recognizing the philosophy of assistance from commercial entities. Even though it built great shops, it did not staff them, and more and more was having its work done by contractors.

KEN: Less hands-on, as with NACA.

CALDWELL: Less and less of that, and it turned out more and more to be a project center where you farm out and let contracts to other people. Max was appalled by that, and I was too. There wasn't much I could say about it, but it was clear that without the hands-on work, the expertise would soon be lost. You'd just become somebody writing contracts. What was the use of us having the facilities there? Why'd they build them in the first place? If you go over there now, there'll be five guys walking around in a great monster shop, because the task isn't to build things there. The great shop over there has a mockup of the space station, for instance. They don't build or test anything in it—it's just got this mockup.

Anyway, as the Shuttle got moving along, my division inherited many of the mechanical systems; the landing gear, the doors that opened, things that perform functions

Initially, the Shuttle design showed a docking port right on the nose. Subsequently, it got shifted to a thing in the payload bay. It was a tunnel and a port in which one would dock the other spacecraft. There was nothing to dock to. Nothing was in the wind to do it, and pretty soon it was dropped as a line item from the program because there was money allocated to that and the money was needed other places, so they dropped it altogether.

I decided to continue working on the docking system with some other guys. We reasoned that the old Gemini system was really no system at all. It was really just something that was whomped together to join it up. The one on Apollo was a sort of male-female arrangement, which is not good because only "males" and "females" can dock. You couldn't have two "females" or two "males." Besides that, the male part occupies the very tunnel that you need to pass through, so whatever you do, you have to dismantle the thing and get it out of the way before, let's say, a man could go through it. It was okay with the lunar module because it used that kind of system. Because nobody had to get back, they didn't mind going to the trouble of taking this whole thing over. But it just wouldn't do later on, so we reasoned that we should design one that was "androgynous." It turned out that we stumbled upon something very useful. We had no real reason to be doing it other than saying, "One of these days spacecraft are going to want to dock."

Mission to Moscow

CALDWELL: We had no sooner made some mockups and demonstrated that it would work and the concept was good than I got a call from Gilruth. I got to his office and Glen Lunney was there. He said, "I've got a surprise. I've been told that you're supposed to go to Moscow and talk to the Russians about some future space programs."

He went on to explain that Nixon, and I think Kosygin, who was their head man at the time, for purely political reasons had said, "Let's cooperate in space." They had decided on nothing in particular, but had the idea of "Let's have dialogue on this thing." I think it was an honest effort at cooperation and joint activity when everything else was going wrong. Headquarters had decided to send Gilruth and a small delegation to Moscow to talk to the Russians about this, and Gilruth said, "You two guys are going with me. I want to get a guy from Marshall as well, because Marshall's feelings will be hurt if we don't get a guy from there," so he asked George Hardy.

He said, "Now that we can go, what are we going to do? Have you guys got any ideas?" Glen said, "Well, we've been doing a lot of work on guidance and navigation and rendezvous and that sort of thing, which is all a very necessary part of the operational part of the activity," and I said, "We've been working on a docking system. It's a new thing. Nobody else is using anything like it." Gilruth said, "Great! That may not be the right thing but it gives us two subjects that we can talk about."

ROBBIE: Do you know why he chose you instead of other people?

CALDWELL: Maybe it was because we weren't busy at the time! *(laughter)* The only person other than me that he could have picked would have been Max, but Max was the head of the whole directorate and this was going to turn out to be a full-time job. So we got prepared and, of course, we had to go to Headquarters a couple of times to check with everything. One other guy, named Arnold Frutkin, was to go with us. He was at Headquarters and was involved with diplomatic matters at NASA. Anyway, the first thing you know, Army Intelligence, the CIA, the State Department all showed up and they were beside themselves. "Oh my God, sending a bunch of dumb engineers form NASA over to Russia. What is happening?" They gave us all these warnings about what was going to happen and they literally scared the hell out of us. We knew it was just a bunch of BS in a way, but just the same, they were horrified. Really I think they were angry because somebody besides them was going.

ROBBIE: They were horrified because they thought you might give away secrets of the space program?

CALDWELL: And we didn't *have* any secrets—at least as far as I knew, we didn't.

Anyway, we went over there and arrived in October 1970. We came in at 4 o'clock in the afternoon. It was dark and snowing lightly. The airplane rolled up to the stop and we looked out the window and there were all these guys in greatcoats with the sleeves hanging down over their arms, and big fur hats on. A bunch of them had submachine guns. I thought, "What have we gotten ourselves into?" And, of course, this was right in the middle of the Vietnam War.

Well, they couldn't have been nicer. The man who met us was Dr. Petrov—a real gentleman, scholar, and academician. He was also a member of the Academy of Science. They took us into Moscow and down one of the main drags there were great billboards showing Uncle Sam with Vietnamese babies and blood running all around, stomping them.

ROBBIE: So you thought you were in the wrong place at the wrong time.

CALDWELL: Yeah, it didn't seem like the place to be, but it was nothing like that. The people were very nice. In fact, they almost seemed as if they were afraid I would grab them and bite their heads off or something. They bent over backwards to do everything for us.

It turned out that we had hit the nail right on the head when we had said "operations in space and rendezvous and docking." I had written a paper that described the system, and Lunney had put together a paper describing what he had done. We drafted a memorandum of understanding that said simply that we would continue to have a dialogue about joint space activity. We didn't name anything. Lunny and a Russian guy named Peter Tischkoff, who had been a cosmonaut, drafted it.

He and Lunney did this, and the rest of had to disappear for a while so we wouldn't bother them. I remember Frutkin and I walked out through some of the streets nearby. We didn't dare go too far, but there was a beautiful avenue, lined with trees and tables, and old men were sitting around playing some sort of game like checkers. As we approached, they'd all sort of put their hands over the things and look around like they were just chatting. I gathered that it was some of game that was frowned upon. Maybe they were gambling and you were not supposed to gamble. They wouldn't know that we were Americans, though a lot of people would know that we were not Russian because our clothes were different.

When we came back, we reported all this, and it wasn't long before the Russians said, "We would like to come see you next summer and talk some more." When they arrived the next summer, they said, "Let's have a joint mission next year." We about fell out of our chairs but they didn't hem or haw about it. We did some fast talking amongst ourselves and agreed that there was no way we could do it in a year. They couldn't either, for that matter. But we agreed to talk about it. It would involve Apollo, probably, and one of their spacecraft, which they didn't name at the time.

ROBBIE: And what was the idea of the mission?

CALDWELL: We would dock with them. We were stunned, but a few more memos were written and agreements made, and we went back the next October, and in the course of the meeting they said, "We've got a docking system we propose," and they put this picture on the wall. It was what we had taken over there earlier. It wasn't the same drawing, but it was the androgynous system and they said, "This is what we think we ought to do."

ROBBIE: Did they say, "We took this from what you brought over?"

CALDWELL: No. They implied that it had just been generated there. Maybe it had been.

They also proposed some things that hit right on what Lunney had proposed—operational activities. It was clear that they wanted to do something and they wanted to cooperate. It was not like what they usually did, trying to keep things from happening. They *wanted* this to happen.

ROBBIE: Would you guess that the Russian scientists knew that this was something that their government would put money into because of the symbolic value, so it was a way of keeping their own program going?

CALDWELL: It's entirely possible. They'd lost the Apollo Moon Race. They had not made it to the Moon and they didn't have anything else going, so they may have been in deep trouble politically. There was no question that they intended to cooperate, though. They didn't put up any roadblocks.

And living there was tough. The people that I saw, who were the engineers and the Academy of Sciences people, were living on the top of the heap and they didn't have very much. Most of them didn't have any automobiles, and if we went to a banquet with a big dinner, there was nothing left on that table when the thing was over. If there was any food left, it went into somebody's pocket.

Anyway, the program went on and eventually I withdrew from it. It began turning into a routine activity and I just gradually got out of it. The Russians continued on with the docking system even after Apollo-Soyuz, and they refined it and made it better and better. There were things that needed to be improved, so they continued on even though they didn't have any finite reason to do so.

By that time, I had resigned from NASA and was working for Space Industries. We essentially put together the initial theory and the development of the thing along with these new developments that the Russians had come up with, and presented what we considered the state of the art of docking systems in a short course.

In the meantime NASA—JSC in-house—had been working on docking systems totally unlike ours, and I probably couldn't have got away with this if I'd been working for NASA because I was essentially saying that the thing NASA was working on wasn't worth a damn. It was terrible! I guess the course was fairly well received. There must have been 50 people who attended the two-day sessions.

I was told later that that short course was responsible for NASA electing to go with the Russian docking system on the Shuttle when they docked with Mir, on the strength of the fact that since both sides had agreed that it was a good thing, at least they should consider meeting one side. It was there and it was working; they had demonstrated it.

Robbie: This was essentially *your* docking system.

Caldwell: Well, people like to say "It's yours," but it takes a lot of effort to make a concept come out and work.

CHAPTER 4

EILENE GALLOWAY
GRAND DAME OF SPACE

BIOGRAPHIC INFORMATION

DR. EILENE GALLOWAY (1906–2009), Honorary Director of the International Institute of Space Law and Trustee Emeritus of the International Academy of Astronautics, retired from the Congressional Research Service, Library of Congress, as Senior Specialist in International Relations (National Security) and continued her career in national and international space activities, which began in Congress where she participated in the House and Senate legislative process leading to the National Aeronautics and Space Act of 1958.

The orbiting of Sputnik on October 4, 1957 caused the Chairman of the Senate Armed Services Committee, Senator Richard B. Russell, to direct her assignments to hearings held by Senator Lyndon B. Johnson, Chairman of the Preparedness Investigating Subcommittee, on the Missile/Satellite Situation. When the Senate established its Special Committee on Space and Astronautics, and later the Standing Committee on Astronautical and Space Sciences, Senator Johnson appointed her Special Consultant. In that capacity, she wrote and edited Senate documents on the variety of issues and problems arising from the uses and exploration of outer space. Among her reports are:

- The final report of the Senate Special Committee on Space and Astronautics, Space Law, and International Cooperation and Organization for Outer Space,
- Analyses of Space Treaties formulated by the United Nations;
- The Treaty on Principles Governing the Activities of States in the Exploration and Use of Outer Space, Including the Moon and Other Celestial Bodies;
- The Agreement on Rescue of Astronauts, the Return of Astronauts, and the Return of Objects Launched into Outer Space;
- The Convention on Registration of Objects Launched into Outer Space.

For the Senate Committee on Commerce, Science, and Transportation, she wrote the document "Agreement Governing the Activities of States on the Moon and Other Celestial Bodies, Parts 1 and 2."

Beginning in 1959, Dr. Galloway was appointed to participate in sessions of the United Nations Committee on the Peaceful Uses of Outer Space, and its Legal Subcommittee. Later, she was designated by the International Astronautical Federation as an Official Observer for the United Nations' outer space sessions in New York, Geneva, and Vienna. She served as consultant on national and international space activities for NASA, the Department of State, the U.S. Civil Service Commission, the Federal Communications Commission, and the Office of Technology Assessment. She also lectured at various educational institutions. She was the author of more than 90 articles and reports on outer space activities. After our interview, she wrote, "I am occupied with the International Space Station, which affords maximum opportunities for interdisciplinary research."

Awards granted to Dr. Galloway include the Andrew G. Haley Gold Medal (1968); Purdue University Testimonial (1974); Library of Congress Commendation for Distinguished Service (1975); NASA's Public Service Award and Gold Medal (1984); American Astronautical Society

Fellow (1986); International Academy of Astronautics Theodore Von Karman Award (1986); Woman in Aerospace Lifetime Achievement Award (1987); International Institute of Space Law Lifetime Achievement Award (1989); Lake Forest College Honorary Doctor of Laws degree (1990); Swarthmore College Honorary Doctor of Laws degree (1992); International Academy of Astronautics Award of Trustee Emeritus (1995); American Institute of Aeronautics and Astronautics Award of Fellow (1996); NASA Johnson Space Center Achievement Award (1997), and the American Astronautical Society's John F. Kennedy Astronautics Award (1998). The book *Space Law: Development and Scope* (1992), edited by N. Jasentuliyama, is dedicated to Eilene Galloway. The American Astronautical Society presented Dr. Galloway with the 1999 John F. Kennedy Astronautics Award at a ceremony in March 1999 at the Goddard Space Center. The citation reads: "To Dr. Eilene Galloway, for forty years in advising Congress on legal and policy aspects of outer space, service to the United Nations Committee on the Peaceful Uses of Outer Space (COPOUS), and the development of international space law."

During the 1990s, she was invited to give papers at and participate in:

- The European Space Agency's International Lunar Workshop (1994). (Paper presented: "Political, "Legal, and Economic Aspects of a Return to the Moon");
- The United Nations' 50th anniversary (1995);
- The First Annual Space Technology Consortium Symposium, Arlington VA (1996). (Paper presented: "International Space Law Perspective.")
- The NASA Advisory Committee on the International Space Station (1996–2000);
- The Seminar on Space Futures and Human Security, Alpbach, Austria (1997) at the invitation of the UN and the Austrian government. (Paper presented: "Space Futures and Human Security")
- NASA Office of Policy and Plans/History Office Symposium,

1997: "Reconsidering Sputnik: Forty Years Since the Soviet Sputnik." Paper presented: "Organizing the U.S. Government for Outer Space 1957–1958."

- The International Astronautical Federation Congress, Turin Italy (1997), celebrating the 30th anniversary of the 1967 treaty. Paper presented: "The United States and the 1967 Treaty on Outer Space."
- NASA Johnson Space Center lecture (1997) on the effects of Sputnik
- UNISPACE III, the Third United Nations Conference on Exploration and Peaceful Uses of Outer Space, July 19–23, 1999, "Using Space to Create Global Solutions."
- International Institute of Space Law workshop on "Law in the 21st Century": Dr. Galloway chaired the first workshop session on "Existing UN Space Treaties: Strengths and Needs."

Biographies in *Who's Who in America*; *Who's Who in the World;* Who's *Who of American Women; Who's Who in Science and Engineering; Who's Who in the East.*

INTERVIEW

This three-hour long taped interview took place in Galloway's home in Washington DC on Nov. 20, 1997. The transcript has been edited by both interviewers and by Dr. Galloway herself for clarity, organization, and flow. Dr. Galloway also added extensive supplementary commentaries and additional information, some of which either appears in italics or parentheses or has been incorporated into the text, and some of which remains in the original 70-page interview, which can be found in full at www.davis-floyd.com. Editorial comments from Robbie Davis-Floyd or Ken Cox appear in brackets.

ROBBIE: Ken and I are both intrigued by the fact that you are a woman who had such a position of responsibility in the 1950s. So we'd like to start with your background. Would you tell us where you were born, where you grew up, and what your childhood was like?

EILENE: I was born in Kansas City, Missouri in 1906. So I'm 91 and a half now! *(laughs)* My mother, who was born in Texas of a pioneer family, was 20 years old. My father, who was 12 years older, came from a prominent family in Monroe, Louisiana. My father worked for the railroads. When I was nine, he deserted us—he vanished overnight. We had no money to pay the rent or to buy food.

ROBBIE: What did you do?

EILENE: My mother was resourceful. She rented out some of our living space, and got a job as a telephone operator. I realized that I would have to make my living and that I could not depend on anyone else for this.

First, I learned to be a dressmaker. By the time I was 14, I had taken enough sewing at school and helped my mother enough that I could sew anything. But when I got to high school, I decided to become a stenographer and then a secretary. All through grade school and high school I had wonderful teachers who made us practice. There wasn't any feeling of pressure—it was just that if you put cloth together accurately, it could be a dress. If you combined the right ingredients, you had the satisfaction of a cake. My Sunday School teacher arranged for a teacher to give me music lessons without charge—if you strike the right notes, you can get the tune you want, you see.

ROBBIE: And then as you got older you worked every summer?

EILENE: Yes, and what made these jobs possible was that in school I took shorthand and typing for two years and my teacher demanded absolute perfection. I still use shorthand at conferences. I still take shorthand at 125 words per minute, and I can type 10 pages [on a typewriter], if I have a good machine, without making an error, or hardly any.

When I was 15 and went to apply for my first job, there was a girl ahead of me, and they said, "Have you had experience?" And she said, "No," and he said they wouldn't take her application. So he said to me, "Have you had experience?" and I said, "Yes," and then I sat down with a form to try to figure out what experience I had had *(laughs)*. I remembered that when Miss Kleeman, who was head of the English department in high school, took a class, she would have a girl sit in her office, because she was like the Dean of Women in the high school. So I put down that I worked for Miss Kleeman in this office. Then the man asked me to take off my hat, and I had lots and lots of gold hair. So he walked around, and he said, "That's good because Mr. Fennel will not take a girl with bobbed hair." It was a stigma to get your hair bobbed at that time. It just wasn't the thing to do!

They didn't ask me to do the shorthand and typing, in which I was prepared to take a test. Mr. Fennel looked at me and saw that I had long hair and I got the job *(laughs)*. That was very funny!

I had a marvelous education at Westport High School in public speaking and debate, learning how to speak and how to study pro and con. I was Captain of my Debate Team during my junior and senior years and we won our debates. We were taught how to speak, to throw your voice so you could be heard. I shall never forget reading my first law when we were studying "Resolved that the Philippine Islands should be granted their immediate independence." The law was based on their being ready, and I realized that all I had to do was prove that the Philippines were not yet ready for independence—that was the negative side.

I was very anxious to go to college and I won a four-year scholarship. It was the first time Washington University had given the scholarship to a girl. When I got there, my scholarship covered only the tuition. I had worked all summer, and I had the money to pay for the room, but I didn't have any money for food. Someone had given me a box of chocolates when I left Kansas City, so I ate those for two days, and then one of the girls said, "You might as well come and eat because they start charging you for the food as soon as you arrive." It was six dollars a week, and I didn't have six dollars. But there was an AD from Dr. Shepard, who was head of the Political Science department, for somebody to do shorthand and typing for the afternoon, and it paid forty dollars a month.

I went to the office and he was leaving, so I rushed after him, and said, "Are you Dr. Shepard?" He said yes, and I said that I wanted to apply for the job, and he said, "Well you're a freshman, and you won't be able to do your homework if you spend four hours in the afternoon." I said, "Oh, I'm sure that I can do it. The fact is that I need money to pay for my board." So I got the job, and the man I would marry, George Barnes Galloway, was doing his MA under Shepard. I met him the first week I was there. I saw him frequently, and typed his MA dissertation. I charged him ten cents for the first page, and five cents for each of the carbons. He married me before his Ph.D. dissertation had to be typed, so it didn't cost him anything! *(laughter)*

ROBBIE: That was a smart move on his part.

EILENE: Yes! He had to finish his Ph.D. in Washington D.C. at the Robert Brookings Graduate School of Government and Economics, so we got married and I gave up two years of that scholarship and came to Washington for one year to be with my husband. Then he finished his Ph.D. and got a job in Philadelphia at the Bureau of Municipal Research. We decided to live in Swarthmore, Pennsylvania so I could go to college. Graduating from college was an obsession for me, perhaps in part because in Washington, I had had the unusual experience of being with a small group of men and women taking their doctorates and that was an education in itself—many of them became well-known later.

After we moved to Swarthmore, I went to the college to enroll in the Political Science Department. The Dean said about 1100 people had applied for only a few places—about 21, I think—and I couldn't get in. I went outside and sat on a bench under a tree and cried and cried. Then I went to the head of the Political Science Department, Dr. Robert Brooks, and asked him if I could study with him for a year until I could get in, and he was so moved by my situation that he appealed to President Aydelotte. Aydelotte was starting the Honors Program, modeled on a British idea, and the teachers in that program—Political Science, History, Economics, and Philosophy—voted me in.

The Dean was very upset because he said the College was limited to 250 boys and 250 girls and they all lived on campus. I would make 251 girls and lived off campus! So he said I could come and be credited and so forth, but he could not print my name in the catalog for my junior year because that would make 251 names! At first he wasn't going to admit me, because he said it was against the rules for the students to be engaged. And I said, "I'm not engaged, I'm already married!" And this also upset him very much *(laughs)*.Then on top of all that I had a baby, David Barnes Galloway, during the spring vacation of my junior year, which was unusual for Swarthmore. Of course, the dean did not know that I would graduate Phi Beta Kappa and with high honors from Swarthmore and be appointed to my first teaching job in the Political Sciences Department—a position I held for two years until we moved to Washington. Or that I would return years later to receive an honorary LLD degree recognizing my outer space career!

ROBBIE: What did you major in?

EILENE: The social sciences. I was in the Honors Program, where you took several different disciplines and learned how to coordinate them. It was Economics, Political Science, Law, and Philosophy.

ROBBIE: And why did you want that particular program? What attracted you to it?

EILENE: I was in Political Science because I had spent two years with Walter Shepard, and my husband was a very famous political scientist. He was an expert on Congress and wrote *Congress at the Crossroads* and a history of the House of Representatives, and books like that.

ROBBIE: And Walter Shepard was also a political scientist?

Eilene: Yes. And this was wonderful training! The Swarthmore honors program was intensive, and you learned how to relate these different disciplines, which was helpful to me all my life, in the work I did for Congress and later on when I was invited to be a member of the International Academy of Astronautics. I was responsible for relating space science and technology to concepts of foreign policy, international cooperation, international relations, organization, and management, and economics. When you look ahead, you can't see that there is logic in the way things are developing, but when you look back, you can see that in fact there was some logic to it.

Robbie: Do you remember the teachers who influenced you the most?

Eilene: Yes. One was Dr. Robert C. Brooks, who was head of Swarthmore's Political Science Department, and another was the historian Fred Manning. I learned a lot from Manning about the law and how to do research in national and international law. We would take a clause in the Constitution—let's say, the commerce clause—and then we would study the cases from the beginning of history up to the present, and learn what happened with each court decision. Then I was required to write a thesis, and I wrote one on the work of Charles Merriam, who was a political scientist. Learning how to do research in law was very important to me later, because when I was at the Library of Congress, in the Congressional Research Service, which was then called Legislative Reference Service, I was a National Defense Analyst.

Robbie: And your husband was supportive of you continuing in college? He didn't want you to stop and just be his wife and raise his kids?

Eilene: No, he was always very supportive and pleased when I accomplished something.

Robbie: How long were you married?

EILENE: 43 years, until his death.

ROBBIE: And how many children did you have altogether?

EILENE: 2 sons. And now I have six grandchildren and four great-grandchildren!

ROBBIE: I'm interested in how you managed, because that was a time of special pressure, when many women would have just dropped out of college to raise their children, so it's wonderful to hear the story of how you did it. You just kept right on going.

EILENE: It is a misconception to assume, as many people do, that women were not concerned with getting professional jobs at that time. This assumption surprises me, because I was always with women who were working at one thing or another. It is just that there is more publicity now about the problems than there was at that time. And young people going into these professional fields now have a harder time getting domestic help than I had when my children were small. The only really hard time was during the Depression when my husband's position was abolished, there was no hope for jobs, and we had to live with his parents until Franklin Roosevelt started the New Deal.

The Library of Congress Years

ROBBIE: Was your first major job in Washington at the Library of Congress?

EILENE: Actually, my first job with the government was with the Federal Emergency Relief Administration where I was in the Adult Education Division. Later I was appointed to the Legislative Reference Service (LRS), which is now called the Congressional Research Service (CRS), and is located in the Library of Congress as the research arm of the Congress. It is set up so that members and committees can get objective reports on their agendas and do not have to depend on lobbyists.

At first, I was editor of what later became *Public Affairs Abstracts*. I had an assignment to work on materials concerned with what we would do after the war. The war had just started, but we were sure we were going to win it. The LRS had an admiral and a general who sat in the front office and answered inquiries about military matters. The admiral would start laughing every time he saw me because I was working on the "post-war," when the war had just begun! And this amused him no end. *(laughing)* That was how I got into national defense. Later, this project, which was originally called *Post-War Abstracts*, became *Public Affairs Abstracts*—a sort of *Reader's Digest* for members of Congress on political and economic, national and international subjects concerned with legislation.

ROBBIE: What fascinated you about defense?

EILENE: Well, it was the role of Congress in the reorganization of the Department of Defense, the military budget, and military history—the plans and the battles. I collected a lot of books and historical materials. I wrote a "History of United States Military Policy on Reserve Forces," from the beginning of the country, with George Washington, up through the Korean War. I worked on the main subjects that Congress was concerned with in defense, such as military manpower, the military budget, military history and strategy. As an LRS employee, you work with committees, figure out what the problem is, suggest witnesses, propose questions for hearings, and write reports.

ROBBIE: And you didn't know where this fascination came from?

EILENE: No, I don't know where it came from except that it was something that I was especially concerned with. And it was very helpful later—it provided the background for my work on the space program. You see, in the spring of 1957, I wrote a report that was published by the House and Senate on "Guided Missiles in Foreign Countries." And anybody who had worked on missiles got called up the very next day after Sputnik went up. I had worked on weapons, everything that was on the agenda of the House and Senate Arms Services Committees.

ROBBIE: What was your official title in this position?

EILENE: I began as a National Defense Analyst and was promoted to Senior Specialist in International Relations and then in brackets [National Security]. I had a job description that sounded like I was Secretary of Defense and Secretary of State both! *(Laughter)* It was really funny—every time I read that I think it's very funny.

I started out in September of 1941, before Pearl Harbor, at the beginning government professional level. I made $2000 a year. The Director of the Legislative Reference Service made $5000 a year, which shows you how much inflation we've had since then. And I said to Mrs. Bachley, who was there, "Oh, I had so wanted to make $2400." She said, "Well don't worry, dear, you will be promoted, and the next grade is $2600." And I thought, "Oh my goodness, that's $200 more than I need!" *(laughs)* I was promoted through the years to the highest research level—GS17. I was never discriminated against because I was a woman and was working in the national defense field mostly occupied by men.

ROBBIE: So the Senator, or Representative, or Committee Chairman would come to you and say, "We need information about this topic. You go research it and find it and type it up and bring it to us?"

EILENE: Yes. The way some of the assignments came up is rather interesting. Once the House Armed Services Committee was holding a hearing, and they had a terrible problem because the Defense Department was calling up men from the reserve forces for the Korean War—men who had already fought in World War II. They had come back and started their families, bought cars, and were settled, and meanwhile other draftable men were not being called up at all—it was a terrible problem of inequity.

Well, when the hearing was over, Congressman Overton Brooks saw me and said, "Eilene, I want you to write a history of the Reserve Forces," because they wanted to know all U.S. policy on the subject. It was a fascinating assignment, going back to the trouble that George Washington had with the Reserves.

ROBBIE: In order to write these reports, you would delve through the Library of Congress historical records, and read original material?

EILENE: Yes, and also we had a bibliographic section, and I could give them an order. In 1958, I gave them an assignment before I went to the United Nations, asking for all of the references on outer space.

KEN: Why did you ask for that at that moment? What drove you to make that request?

EILENE: I had to keep track of everything in case I'm asked something. *(laughs)* So I had this stack of three by five cards on everything that came in on space. When I got back from the U.N., where I had gone with Lyndon Johnson, I went through these cards, and I found this many *(holds up fingers about 2 inches apart)* that were on space law. I had never heard of space law. I started going through these cards, and I thought, well, this would make an interesting committee print, and that is one of the Senate documents that I did. (1) [*Editors' note:* The names of the documents associated with these reference numbers appear at the end of this interview.]

So I said to Lyndon Johnson, "I think I could get out a committee print on space law." And he said, "Well, Eilene, if you want to do that then go ahead, that's all right." He was sort of offhand. It was late in November and we only had one month left of that Congress. So I got all of these materials together, and the document was published on December 31, 1958. It was the first thing that came out on space law and within two weeks, every copy was gone, there was such a demand for it. Everybody who was working on aviation, aeronautics, sovereignty wanted it. So then I had to get this study out again, and instead of calling it "Space Law," which some people thought was esoteric, I called it *Legal Problems of Space Exploration*. The next time, I called it *Legal Problems of Exploration and the Uses of Outer Space*. So every once in a while, the committee would get an update on space law, so you have a record of the national and international law that regulates space activity.

Robbie: You were working for the Library of Congress in 1941, and you were being asked to do all these reports, and I imagine one of the things that sparked your interest in defense was the war. Were you already interested in defense before Pearl Harbor, or did Pearl Harbor galvanize your research interest?

Eilene: National Security was also involved with international relations. But Congress had a role in the organization of the military departments and that is part of political science: government organization and management. And I was actually *interested* in the battles, the war plans and strategy and everything about it, because assignments were opportunities for interdisciplinary research. That's why Congressional inquiries about matters of national defense were assigned to me, so I could relate them to foreign relations. When we got the 1967 treaty on principles guiding nations to conduct their space activities, I was working for the Senate Space Committee, and I was also working for the Senate Foreign Relations Committee. Some of the same Senators were on these committees, so you could combine them—you could combine the whole thing—the authorization, the appropriation, and whatever the committee was supposed to do. Of course, it spilled over into a lot of other subjects.

Robbie: Did your work change dramatically after the war ended?

Eilene: No, I still worked at the Congressional Research Service until I retired, but I was practically delegated to some of the committees. When Lyndon asked me to work for him on outer space, the

Washington Post came and my husband opened it, and here was a picture of me, and an announcement that Lyndon Johnson had appointed me to be a Special Consultant to the Senate Special Committee on Space and Astronautics.

This was while we were working on the NASA act in 1958. I worked on the NASA act all through the House and Senate hearings. Johnson didn't make a request to the Head of the Congressional Research Service, and he didn't say anything to me—I had never seen this picture before. He just appointed me.

KEN: When the newspaper came, it was a surprise to you?

EILENE: It was a surprise to all of us! *(laughter)* Jonathan wondered why I was a "Special" Consultant," and I said, "Well, it is a 'Special' Committee, I suppose it was a typographical error." I couldn't think of any other reason. Usually people appoint you as just a consultant. That was how I came to work on organizing NASA.

ROBBIE: And the "Special" was not a typo?

EILENE: No! *(laughs)* I was also working for Congressman McCormack, who was Chairman of the House Select Committee on Astronautics and Space Exploration. These hearings were held in April and May 1958 on H.R. 11881 about creating NASA. Congressman McCormack began the hearings with the message from President Eisenhower and the report he had asked me to write on "The Problems of Congress in Formulating Outer Space Legislation." McCormack began the hearings with a message from President Eisenhower, and my report. This was a wonderful experience because it enabled me to get more information before Lyndon Johnson began his second set of hearings in the Senate. (2)

You will recall that soon after the Sputnik was orbited, Lyndon Johnson, as Chairman of the Preparedness Investigating Subcommittee of the Senate Armed Services Committee, began extensive hearings on

"Inquiry into Satellite and Missile Programs." These hearings began on November 25, 1958 and were continued through three volumes that came to 2470 pages. Part 3 includes February and April and July 1958. Lyndon Johnson's specific hearings on the NASA legislation were held by the Senate Special Committee on Space and Astronautics during the 85th Congress, second session on S. 3609. Two parts were held in May 1958, coming to 413 pages. (3)

The reason we had this Special Committee was that the jurisdiction of the Armed Services Committee was over the Department of Defense and military matters, and we were engaged in setting up a civilian agency over which the Defense Department would not have control. We were dealing with civilian benefits as well as guarding national defense aspects. I helped with analytical reports and formulating questions for the hearings. The subject could not be continued by the Senate Armed Services, because they deal only with defense. And we had a lot of special benefits that had to be attended to. The appropriations for defense would not have permitted them to develop global communications and meteorology and all these things that we have developed for civilian space—that's why we had to set up NASA. So in order to do that, they had this Special Committee in the Senate, and this Select Committee in the House, and I worked for both of them and was available to these committees all the time.

KEN: Was there anything you had worked on before Sputnik, not knowing it was going to happen, that fed into this?

EILENE: Well, my report on "Guided Missiles in Foreign Countries" was published in the spring of 1957.

ROBBIE: Between 1945 when the war ended and 1957, you continued working for the Congressional Research Service doing whatever reports you were asked to do?

EILENE: And they were all on national defense, and of course, the atomic bomb went off in 1945. The Legislative Reference Service published reports called *Public Affairs Bulletins*. I didn't know anything about atomic energy, but I went to Dr. Griffith, the director, and I said, "Could I write a *Public Affairs Bulletin* on atomic energy?" and he was so happy that somebody was willing to do this. No one else wanted to do it. I knew nothing about it, but I got the Smyth Report and read it three times—I thought my brain would crack trying to get this technology in my head. And I wrote a report called "Atomic Power Issues before Congress," and then I waited to see what would happen *(laughs)*.

Major General Groves, who was head of the Manhattan Project for Atomic Energy, sent for copies, and then the Navy sent for copies and used it for a textbook in one of their classes. In 1947, I was co-author with Bernard Brodie of a *Public Affairs Bulletin* on "The Atomic Bomb and the Armed Services." I worked on legislation related to the Joint Committee on Atomic Energy—Senator McMahon was Chairman of the Committee working on this legislation.

ROBBIE: What was the legislation about?

EILENE: It was about how to organize government to take care of atomic energy, which could be used for peace and war. Congress created a joint committee. This was what carried over to outer space matters, because after the Sputnik went up and we had to organize the government for space, we also had to organize the Congress, and there was a lot of confusion about this. That was why McCormack asked me to write this report on what the problem was. In addition, we had four options for organizing the Congress. Space cut across the jurisdictions of so many committees and so many agencies that you wouldn't know where to send legislative proposals for a budget or a program.

So we considered whether to have a joint committee because there was a strong feeling about giving space to the Joint Committee on Atomic Energy. Well, this would have been a disaster because atomic energy, while it could be used for peace and war, similar to space technology, was a source of energy, and outer space was a place where we were doing many things that were civilian and had peaceful uses. The whole objective finally became how to organize the government to carry on a space program for the benefit of all mankind, and avoid orbiting weapons of mass destruction.

Robbie: How did that become an objective?

Eilene: Because of the testimony of the scientists and the engineers, who told us about all these benefits that we could develop from using outer space. So when it came to analyzing the atomic energy option, I had to prove that we should not be giving jurisdiction to the Joint Committee on Atomic Energy.

Ken: Because there were groups that thought that was the right answer?

Eilene: The whole subject was somewhat amorphous then, and neither the House nor the Senate members wanted to set up a Joint Committee on Outer Space. The House people felt overshadowed by the Senators on a Joint Committee. The other option was to give legislation to the existing committees, but that would mean cutting up the subject, and it wouldn't be unified in one place. The other option was to set up separate committees in the House and Senate, and that was what we did. First we had two separate House Select and Senate Special Committees, and then after we got the NASA act, we had permanent committees. It was very unusual for the Congress to set up two new permanent committees.

Robbie: When you wrote, you must have known what Congress *ought* to do. Do you think your report helped shape what Congress *did* do? Were you able to influence what Congress actually did?

Eilene: When you work for the Congressional Research Service, you are supposed to do objective research, and you're not supposed to tell them what to do or what not to do. But (*laughing*) if someone asked me, "What do you think about it?" I would tell them what I thought about it in a way that left the decision up to them.

You're supposed to write a pro and con report. Once I wrote a report on the military draft. I wrote the pro and con on that, and you could read it and be paralyzed and not know what to do, because it would be so evenly balanced. But some subjects were not like that. Once you have explained why you shouldn't give legislative jurisdiction of outer space to the Joint Committee on Atomic Energy, the decision becomes almost obvious.

ROBBIE: So your briefs were very influential in setting policy?

EILENE: Well, I hope so. I know that they were in certain cases. But sometimes people would just ask me. One time, a new Senator came in who had been appointed to the Senate Space committee, and he called me over and said, "Now, Eilene"—everybody calls everybody by their first name on the Hill—"I've been put on this committee, and I don't know anything about it, and I want you to keep me from making a fool of myself." *(laughs)* I said I would try!

ROBBIE: World War II had ended not long before this and Korea had ended very recently. The Cold War was under way and it would not have been illogical for the U.S. policy on space to be possessive. The United States could have said, "We're going to be dominant in space, and we're going to use space for defense." Why didn't it happen that way? Why was space, from the beginning, thought of as an internationally cooperative place, rather than as a place where the U.S. should exhibit its military dominance and rule the world?

EILENE: Space science and technology are inextricably international. The satellite goes around the whole Earth in about ninety minutes—it pays no attention to boundary lines. We have to have tracking stations all over the world. The whole point is to get information up and down. You couldn't handle it unless it was international. You have to have arrangements that take advantage of the facts of space science and technology. Development *has* to be international. In the United Nations, our policy was to get some kind of international agreement to ensure that space would be used for peaceful purposes. The technology itself is essentially international, and you can't make it national.

Ken: What were some of the technology and science backgrounds that you said came forth and had a heavy influence on what was happening?

Eilene: I explained that in the Senate document on "International Cooperation and Organization for Outer Space" between 1957 and 1965. In 1967, we had arrangements with 84 countries for international cooperation. And these were the categories: experiments on national science, sounding rockets, meteorological satellites, communication, ionosphere, solar eclipse experiments, scientific satellites, manned flights, deep space, optical, Moon launch, data acquisition. This list grew as the years went on.

Ken: Right before the Space Act in 1958, what were the organizations that were highly influential?

Eilene: The main one was the International Geophysical Year. The IGY was planning to do research on the Earth, oceans, atmosphere, and outer space from July 31, 1957 to the end of 1958 during peak sunspot activity. They were highly organized, in the U.S. especially. Lloyd Berkner, at an informal meeting near Washington, had suggested organizing another polar year. There had been two before, in the 19th century and in 1932. They were organized according to their disciplines: physics, chemistry and so forth. They were also organized geographically, according to countries. Then they were organized into regions and internationally. The scientists and engineers were already planning to study the earth, the atmosphere, and the ocean, and they added outer space.

Robbie: Why did they add outer space?

EILENE: Because that was the main thing coming up. Rockets were being developed as the tools to do all this research.

ROBBIE: So they added outer space because the technology existed to include it.

EILENE: The IGY was working on this and they were organized in connection with the National Academy of Sciences, and they had input into the National Science Foundation. The National Science Foundation was able to get from Congress $43.5 million for our part in it. In addition, we furnished material—the Navy supplied ships, for example. And that's how the Navy happened to be working on Vanguard. That was a purely scientific undertaking. It was separate from what was happening in the National Advisory Committee on Aeronautics, which was also doing scientific research, and the Department of Defense.

When Sputnik first went up, this was perceived as a problem that was critical for U.S. national defense. It was evident that the USSR had the capability to launch intercontinental ballistic missiles. Their first satellite weighed 184 pounds. And then on November 3rd, the next satellite weighed 1120 pounds, so that was why we had these hearings that are recorded in the Senate documents—U.S. pre-eminence in defense, science and technology was facing a direct challenge. (4)

ROBBIE: Were you present at all these hearings?

EILENE: Yes. I sat behind the senators so the chairman could get in touch with me.

ROBBIE: What would your job be after the hearings?

EILENE: I was asked to suggest questions for the witnesses. I did not write the report—it was written up the following January and was just like a press release. But all of this was to try to find out what our situation was with regard to satellites and missiles because everybody was so scared. I assisted Senators Caldwell [D, TX], Symington [D, MO] and Stennis [D, MS]. I did a lot of work for Senator Symington on the international aspects and getting Section 205 into the NASA act. They didn't have Section 205 when Eisenhower sent the bill over.

ROBBIE: What does Section 205 say?

EILENE: It authorizes a program of international cooperation with nations and groups of nations. This was in the preamble of the Eisenhower proposal but it wasn't in the latter part of the text.

ROBBIE: I'd like you to tell us the story of what happened after the launch of Sputnik from a personal point of view. What immediate difference did the launch of Sputnik make in your life? What were the actions that you took as a result? Do you remember where you were when you first heard about Sputnik?

EILENE: Yes, I was at the library. I had an office on the top floor of the annex building. News came through on a late Friday evening and early Saturday morning. I think I must have heard it on the radio, and then I heard it from Senator Russell, who was Chairman of the Senate Armed Services Committee. He telephoned and asked me to write a report on the effect on the United States of the Soviet Union being first to orbit the Sputnik. And then he told Lyndon Johnson that I would be of help, so that was when Johnson telephoned, and I started working on these hearings. They called me because I had written the report on "Guided Missiles in Foreign Countries" and had done other reports for the Senate Armed Services Committee on organization of the Defense Department, military manpower, and questions for hearings with the Joint Chiefs of Staff. The first thing I had to do was write that report for Senator Russell on the impact on the United States of the Soviet Union being the first to orbit the Sputnik. Everybody was scared to death because they thought that the Soviet Union now had the capability of launching intercontinental ballistic missiles. So hearings got started right away. Sputnik was launched October 4, 1957, and the first hearing was November 25, 1957. (5)

ROBBIE: Can you summarize what happened?

EILENE: They concluded that the Soviet Union was first in space, and that the United States would have to put on a special program and become preeminent in space, and this needed to be done immediately, which meant that they had to get organized. And in the meantime, they passed two pieces of legislation to let the Department of Defense go ahead with what they were doing, until we got the NASA act.

ROBBIE: What was the Department of Defense doing?

EILENE: They were working on the Vanguard, and the Army was working on the Redstone missiles, which finally put up the Explorer.

ROBBIE: Vanguard was part of IGY?

EILENE: Yes. Remember that the IGY was a scientific program to study the Earth from the point of view of the land and the oceans and the air and outer space. So that was what this was all about—in the course of these hearings, in that November, very soon after the Sputnik, at a time when we thought that this was critical for U.S. national defense—and that was the problem. The scientist and engineers came up and told us about all these benefits that we could get from space communications, meteorology, navigation, and remote sensing. So the problem then became a bigger problem. It wasn't only national defense and the prevention of weapons of mass destruction from orbiting, it was also the hope and opportunity of developing the future of outer space for civilian uses. That meant that we had to decide on setting up a civilian agency. So we had a better idea about what the problem was and what we had to do as a result of those hearings. That was why the Congress set up the House and Senate Special and Select committees to deal with the NASA act.

ROBBIE: And when you wrote these reports on guided missiles and rockets, were you working just from information available to you in the library, or did you call up people who were actually developing the guided missiles and interview them?

EILENE: I had to work on materials that were in the library on the guided missile study. When the Congressman first called me about this, I said, "Well, you know, most of that information is classified." He said "Yes, but as a member of the Intelligence Committee, I know so much about it that I can't talk about it. If you write the report from unclassified sources, I will be able to use that to talk about the subject publicly."

A lot of this material was written in a number of different languages. We have a lot of people who know these different languages on the staff of the library, so I reached out to get these people, because I had stacks of material with information on missiles in German, French, and other languages. And I remember staffers who sat by my desk and translated orally into English and I took it down in shorthand. After the report came out, a Colonel came from the Strategic Air Command, and he looked at me rather severely, and said, "Now Mrs. Galloway, you have a target in here that we don't have, and we want to know how you got it." And you would think that I had been to Siberia and walked around and found it. It was very funny! *(laughter)*

ROBBIE: What did you say?

EILENE: I said, "Well, I got that from a travel magazine that is published in Australia." It was open information. You almost got so you could put two and two together where some of these things were. But anyway, I had the source, I didn't have to go to Siberia and find it! And then I got acquainted with someone who worked in the law library who knew German, French, and English. So I got him an assignment to do a study on the national laws that had to do with missiles and defense, and he was inspired. He was just thrilled to death with the assignment! This was an entirely different part of the library from the one I was in.

So the director called me in and asked me how I was getting this person who was not on our payroll to do this work *(laughs)*. And I said, "Well, I met him in the snack bar, and he was bored, and I needed these foreign national laws." His report was published as a committee print—the Senate published it—and it was a wonderful help, if you were dealing with a foreign country, to know the aviation laws that might be related to space.

ROBBIE: When did people begin to discuss the formation of NASA?

EILENE: The Rocket and Satellite Research Panel had been working on outer space for ten years. In November 1957, they sent their report on "A National Mission to Explore Outer Space" to the President and Congress. The American Rocket Society also proposed a civilian "National Space Establishment" to be separate from military space. Right after Sputnik, the scientists and engineers proposed establishing a civilian space agency separate from the Department of Defense, and it was a very hot issue, because some of the people in the Department of Defense wanted to keep what they were doing. Ninety percent of our space work was in the Department of Defense. And naturally, they wanted to keep it, yet they could not develop all these peaceful civilian uses. We couldn't expect that they would be using a defense budget to develop global space communications, for example. We got the NASA act in 1958, and by 1962, we had the Communications Satellite Act, whereby the communications satellites developed by NASA were turned over to private industry. And we got the Comsat Corporation, which ultimately led into the Intelsat. There are now 119 nations that are members.

ROBBIE: Was there a big turf war between the Department of Defense and those who were advocating a civilian space agency?

EILENE: Only in the early discussions, when officials in the Department of Defense wanted to keep outer space developments in their area. It was obvious that DOD could not develop the variety of peaceful uses of space that were possible, such as the communications systems I just mentioned. The issue was definitely decided by President Eisenhower, who sent Congress the bill to establish NASA. It showed the power of the engineers and scientists, when they were well organized, to affect policy. We had to ensure that space science and technology would be used for beneficial purposes, and now we know that this was the right decision because we have had forty years of international space cooperation for beneficial purposes, as well as spectacular advances in exploration.

Robbie: Can you name some of the scientists and engineers who were involved?

Eilene: Well, one was James Van Allen, who had an experiment on the Explorer satellite that discovered the radiation belts that are named for him. Another was George Sutton, who was President of the American Rocket Society. Then there were Fred Whipple, Kraft Ehricke, J. Kaplan, Homer Newell, William Pickering—there were almost thirty.

Robbie: Do you remember any of the highlights of their reports and testimony?

Eilene: They wanted the United States to be a leader in using space science and technology for space exploration. They were unanimous in deciding that all the potential uses of space could not be developed by a department whose jurisdiction was limited to military uses, but they realized that there must be cooperation between civilian and military organizations. They also identified goals for exploration, including sending men to the Moon and back, and even establishing a colony on the Moon. Of course, the main peaceful use was development of a global space communications system.

Ken: Who can you name who was key within the Eisenhower Executive Department who supported these activities?

EILENE: Killian was one, and people working in the old National Advisory Committee on Aeronautics, and Paul Dembling was active. Then there were people in the Bureau of the Budget who were working on the civilian aspects.

ROBBIE: After these hearings, what was the next step?

EILENE: We couldn't go forward with this committee anymore, because it was all defense. So we set up the two other committees I've already described—the Senate Special Committee on Space and Astronautics, and the House Select Committee on Astronautics and Space Exploration. These committees then worked on getting the NASA act through. And after the NASA act was through, then we set up the permanent committees.

ROBBIE: Do you remember who first came up with the name NASA?

EILENE: I guess that it evolved from the NACA (National Advisory Committee on Aeronautics).

ROBBIE: At some point someone must have said, "National Aeronautics and Space Agency."

EILENE: You could ask Paul Dembling that. [*Editors' note:* We later did: see Dembling interview below]. I was in Congressman McCormack's office when he was asking me to do this study on the problems of Congress in formulating outer space legislation, and he asked me something about NASA—he had the bill that Eisenhower had sent over to set up the National Aeronautics and Space Agency—and I said, "I really don't like it called an agency! We have so many agencies, and this is such a very special thing, and it affects so many of the other agencies." "But," he said, "what should we do? We've already started calling it NASA." And I said, "Well, we could call it an Administration and have an Administrator, and that would give them more status." Well, he rang a bell on his desk, somebody came in, he said, "Take this bill and take out 'Agency' and 'Director' and in every place put 'Administration' and 'Administrator.'" I was *astonished* by this! *(laughing).* I just happened to be in the right place at the right time! But it did make a difference.

ROBBIE: Can you pinpoint the difference it made?

EILENE: Well instead of being just another agency, NASA was different. It was higher—it carried what I wanted it to carry—it was above all these agencies that had to be coordinated. It gave NASA a coordinating function.

KEN: That *is* fascinating!

EILENE: Incidentally, when Glennan came in as the first administrator of NASA, Dryden was the man under him who had been in charge of the NACA. And I had a great respect for Dryden because his experience had been with national science and with aeronautics, and he expanded into the international field wonderfully well.

ROBBIE: You sat in on all these hearings, and then you had to write a final report.

EILENE: Yes, I wrote the final report. I have a copy that gives the whole history; everything about the two committees—what we decided, who was on it, and everything. (6)

ROBBIE: What happened after that?

EILENE: The committees that were established were authorizing committees, not the appropriation committees, but Lyndon Johnson worked it out so that NASA's budget had to be authorized every single year. At first they objected to that, because the scientists always say they want their money. Their attitude is "Give me the money and leave me alone." *(laughs)* They don't want to be bothered. But it was very important to authorize NASA's budget every single year, because it meant that members of the House and Senate on those committees were becoming experts on the subject, and could help them a lot. So they finally realized that it was a very valuable action.

Then I worked on space law, and of course, the most exciting thing was when Eisenhower asked Lyndon to go to the United Nations to urge creation of the Ad Hoc Committee on the Peaceful Uses of Outer Space, and he was successful. After the NASA act was passed, here was Eisenhower, a Republican president, who had not been much interested in space when Sputnik first went up, and Lyndon was the main force getting all this done. He was like a dynamo in getting all this together at a time in the beginning when Eisenhower was very lukewarm about it. Eisenhower didn't understand the impact on the public, not only here, but worldwide. Sputnik was one of those cataclysmic psychological events that made people understand how important this was.

Lyndon Johnson was the person who was putting all this through. I should say about the NASA act that I had an effect on it because I wrote the report that they needed to establish the National Aeronautics and Space Council. That was not in the bill when it came up to the Hill. It seemed to me that if we divided jurisdictions between NASA and DOD and could anticipate other agencies with space-related functions, like Commerce, Interior, and the State Department, it would be necessary to coordinate all of these civilian agencies.

So I wrote a staff report on the need for a coordinating function of the Council, and that is in the final report. Lyndon Johnson liked the Council idea, and it was very curious the way he got it adopted. The House had first thought they wanted a Joint Committee, then they voted for their separate committee, before the Senate took action. After the House had decided they were going to have a separate committee, the Senate committee met and Lyndon put through a Joint Committee. So you had the Senate passing a Joint Committee and the House having one committee. And Lyndon did this so that when we got into conference, he could say, "Well, I will agree to having the two separate committees, if you agree to the Space Council."

ROBBIE: Who did you say should be on the Space Council? What form did you recommend that it take?

EILENE: The President was in charge of it, and members were the Secretaries of State and of Defense, the NASA Administrator, the Chairman of the Atomic Energy Commission, one other department head, and three distinguished experts from outside the government. When Kennedy was elected, instead of the President being Chairman of the Space Council, the Vice President, Lyndon Johnson, became Chairman. Later on, Nixon abolished the Space Council, and it wasn't because there was something wrong with it, it was because they were trying to get rid of too many functions that were in the White House.

I had another effect on the NASA act *(laughs)*—I was at a meeting of a little group with Senator Symington, going over the text of the bill, which stated that there should be no duplication between the Department of Defense and NASA. And I knew that if both of them had weather programs for their own special purposes, which they needed, there might be some duplication. So I waited until my turn came to talk—I didn't want anyone to say I was a woman trying to dominate the conversation—and when it came my turn, I said, "I think we should put in the word 'unnecessary'." And Senator Symington said, "Eilene, what do you mean by 'unnecessary duplication'?"

And I knew that I didn't have much time to explain it, and I said, "Well, when you are building a house, and you put on a roof, and you have duplicate size shingles and they overlap, it keeps out more rain." So everybody laughed, and Senator Symington said, "Well, put in Eilene's 'unnecessary'." So that is in the NASA Act. *(laughter)*

Then on Section 205 on international cooperation, I worked with Senator Symington to get that in, and there was a great resistance on the part of the State Department. They wanted to have this international cooperation program under their control. Also, there was one very strange thing—the bill said that NASA, in certain circumstances, "may act on behalf of the Department of Defense." We couldn't have that. NASA would not have had any clout over the DOD. There was no way that NASA could do that, so we took that out.

We had a meeting of all the Senators, with the staff sitting on the sidelines, and the Bill stated that the United States would carry on a program of international cooperation with nations and groups of nations, under the supervision of the President of the United States. We had trouble with the State Department on that—they said it was unnecessary, but we put it in anyway. Senator Green was sitting there—he was a very old man, and he was Chairman of the Foreign Relations Committee, and I thought he was asleep, but all of a sudden he woke up and said, "by and with the advice and consent of the Senate." And I was *distraught* because the Senators were about to adjourn and I knew that NASA couldn't carry on a full blown program of international cooperation if everything had to be in the form of a treaty. That was disastrous!

So I spoke with Lyndon Johnson and Herbert Reis—who was in the legal part of the Department of State—to see what could be done. The result was that when Eisenhower signed the bill, he stated that he regarded *"this section merely as recognizing that international treaties may be made in this field and as not precluding, in appropriate cases, less formal arrangements for cooperation. To construe the section otherwise would raise substantial constitutional questions."* This was probably the most significant action I ever took in my professional life.

ROBBIE: What do you mean by "significant"?

EILENE: It meant that NASA was given the scope required for an international space program—it was not restricted by an unnecessary legal barrier. We had less formal arrangements than treaties, such as MOUs [memoranda of understanding], letter agreements, executive agreements, and they permit NASA to carry on a big and diversified program, which is needed. You can see that they couldn't carry out that policy of international cooperation if everything had to be in the form of a treaty!

Several years later, after some experience with international agreements, Senator Symington and Senator Margaret Chase Smith began to worry that NASA was doing too much in this way, and they wondered what NASA was doing—maybe they were doing things that *should* be in the form of a treaty.

So I went down to NASA in Washington, DC. And I got all these bilateral and multilateral contracts that NASA had, and you can see when you look at them that they would not require a treaty. If all you are doing is having a memorandum of understanding about a tracking station, for example, or you want to make an agreement about weather balloons, you don't need a treaty! I prepared a Senate document that was published by the Senate Committee on Aeronautical and Space Sciences in the first session of the 89th Congress. (7)

ROBBIE: With the contracts you gathered at NASA, which you included in this document Number 44, you could show that a treaty was not the only form an international space agreement could take—that indeed, in such cases, a treaty would be completely inappropriate?

EILENE: Yes! Anybody who looked at the document could tell. Yet it was important to be able to do all these things. We have MOUs with ESA (European Space Agency), Canada, Japan, and Russia. We need to have a comprehensive program on science and engineering. So I think I had some effect on the NASA act—I was very pleased to get that straightened out.

ROBBIE: You know, it seems to me as I have listened to everything you've said, that you played a very motherly role in birthing NASA. Why did you care so much about it?

EILENE: Well, because I was so much in favor of international cooperation and peace, and you see, by the time we got through these hearing documents, the objective was to set up a system of peaceful uses for the benefit of all mankind. And that's the way the NASA act begins: Congress declared that "it is the policy of the United States that activities in space should be devoted to peaceful purposes for the benefit of all mankind."

Never a Dull Moment: Working for Lyndon Johnson

EILENE: Let me tell you about one of the interesting episodes with Lyndon Johnson. He always liked to have a unanimous vote on a report. So I wrote the "Final Report of the Senate Committee on Aeronautical and Space Sciences," and he said, "Get a unanimous vote on it." So I went around to all the Senators who were committee members and got all the votes, except for one Senator.

This Senator was abroad—his wife didn't know where he was, his secretary didn't know where he was, no one knew where he was at the moment—and I was frantic, because when you're doing something with Lyndon Johnson, he wants it done *instantly*. So I went back to my office and I wondered what I was going to do. They said the Senator had been in Italy, and they thought he was going to Spain. I thought, "Well, the Navy is good at communication, they have ships in the Mediterranean, I'll call the Navy" *(laughing)*. I called the Navy and explained that *Lyndon Johnson* wanted to get in touch with this Senator to get his vote. They located him. He was on a plane flying from Rome to Madrid when he got this call on the plane. He was astonished, and so was I! I told him all about this report, he said, well, he liked it all except for this one sentence, and I got so worried, I thought, "If I don't get his vote, I'll have to go back to all these other Senators." And I said, "Well, Senator, if I took out the 'and' and just put in a comma, would the comma make you happy?" And he started to laugh, and he said, "Well, it never occurred to me that a comma could make me happy, but," he said, "I think that it would." And I said, "Oh, thank you, thank you!" *(laughter)*

You know, when I was small, I read a lot. We didn't have a television or radio, and I read poetry and classics, like Dickens and all the English novels, things like that. I spent a lot of time in the local library reading books. I read all the fairy tales and they influenced my life's philosophy. The princess was thrown with a bale of straw into a dungeon and told to weave it into gold overnight or she would lose her head or suffer some such misfortune. However, she *always* seemed to do what was expected, met a prince, and lived happily ever after. An impossible task with an impossible deadline—what better philosophy for working on outer space for Lyndon Johnson!

ROBBIE: Johnson was a real Texas cowboy in many ways!

EILENE: Well, at this time I was invited to the first Space Law Conference that they had in Amsterdam and The Hague in 1958. I was invited to write a paper, and I wrote a paper called "The Community of Law and Science." But I didn't expect to be able to go. So you can imagine my astonishment one afternoon—I was sitting at my desk in the middle of the afternoon when I got a call from Lyndon Johnson's Staff Director, saying that I should go to this conference—the President had told him, "Get Eilene on that plane!" So my husband drove me to the State Department—I only had an hour to get a special passport—I went home, threw everything on the bed, shoved it in a suitcase. The Air Force came and drove me to Andrews, and I got on the plane.

There were six Senators who wanted to go to this conference, but Johnson wouldn't let them go because he needed their votes on some Civil Rights legislation, and he said, "Well, Eilene doesn't vote in the Senate, she's expendable, we'll send her." So because I was expendable, I got to fly on this official Air Force plane. In the middle of the night, I was awakened because the plane had landed. I thought I was in Newfoundland, but I was told I was in the Azores, which surprised me. When you did anything for Lyndon Johnson, you had to be ready for the unexpected. The pilot said, "Well, we're going to stop here because they are having refreshments in the Officer's Club." And I said, "I'm so sleepy, I think I will just stay here and not get out." He looked very distressed and he said, "Ma'am, in order to fly on this plane, you have been given the status of a Four-Star General, and nobody can get off until you get off!" *(laughter)* So I woke up instantly—I'd never been a Four-Star General before—and I went and had the refreshments and got back on the plane, and I stayed awake all the way to Orly to enjoy my tenure as a Four-Star General!

ROBBIE: When Johnson became President, were you still working for him at that time?

EILENE: No, I was still at the Congressional Research Service. I never left the Congressional Research Service because I had permanent tenure—I had the highest grade you could get for research in the government, and I was very happy with what I was doing. There was one time when President Johnson wanted me to join the Senate staff, and I had to argue that I could do better where I was in the Library, because I didn't want to give up my permanent tenure. That went back to my childhood experience with economic insecurity—you don't have any permanency on a Congressional staff. If a Senator dies or gets an assignment on another committee, you are out of your job.

Space and the United Nations

EILENE: I should tell you more about President Eisenhower asking Lyndon Johnson in November of 1958 to go to the United Nations, because that is a very dramatic episode that's interesting for American history in terms of Republicans and Democrats, and the relation between the Executive Branch and the Congress, in addition to this subject, which happens to be about space.

Eisenhower had put a resolution on the agenda of the United Nations to set up an AD hoc committee on the peaceful uses of outer space. This committee was committed to peaceful uses for the benefit of mankind, and to avoiding weapons of mass destruction. Eisenhower invited Lyndon Johnson to go to the United Nations and give a speech in favor of setting up this committee in order to demonstrate the unity between the Executive Branch and the Congress, and the Republicans and the Democrats, because he didn't want other nations to think there was going to be an impasse on this issue.

At that time, I happened to be in San Antonio, Texas, attending a Space Medicine conference. So Lyndon Johnson came with some other staff members to where I was, and we all went out to Johnson's ranch to work on the speech. I had not been there before, and the place astonished me. There was a telephone every few feet around the swimming pool, there were telephones up and down the halls, outside any room, any place you went, you didn't have to walk far to find a telephone. We discussed what should be in this speech, and then we met in Austin to go over it. I didn't know who had written the first draft, but Johnson passed it around to see what we thought of it, and I was appalled, because it didn't really say very much of anything.

So, he went around the table, and when he got to me, I didn't say anything for a while. I didn't know what to say, and he said, "Eilene, you get out from under the table and tell me what you think of this." And I said, "Well, if you're going to the United Nations, you have to give a statesmanlike speech at a very high level, and I think this won't do." At that point *(laughing)* I was embarrassed, but Lyndon Johnson was very good at breaking somebody down and getting them to tell the truth.

We all went back to the ranch to work on the speech some more, and then Eisenhower sent down the plane, and we all got on the plane and flew to La Guardia, where we were met by Henry Cabot Lodge, who was our ambassador to the UN. Lyndon Johnson gave his speech, we got 18 other nations to co-sponsor, and the United Nations created the Ad Hoc Committee on Peaceful Uses of Outer Space, with really wonderful terms of reference on what it was designed to do.

There were some countries that would not go in with us, such as the Soviet Union, Czechoslovakia, Poland, the United Arab Republic, and India. They resisted making decisions by majority vote. Naturally the Soviet Union didn't want to have a committee with Morocco and India and other non-spacefaring countries making decisions that affected their space program. After about a year, we worked out the problem by agreeing to make decisions by consensus. That meant that every member nation had to agree before we could go ahead with it, and then the Soviet Union joined, and we got the committee started.

It began with 24 members, and now has 61 members, and the United Nations has an Office of Outer Space Affairs. It was located in New York, but now is in Vienna, Austria. This is the committee that has worked out five space treaties.

The first one was the 1967 treaty, which is the *Magna Carta*—the mother treaty that regulates everything. This is a treaty of general principles adopted by 93 nations on how they should conduct their space activities, because it's obviously an area that has to be regulated. That's another reason it's international—space activities must be regulated, especially Space Communications, which is regulated by the International Telecommunication Union, because geostationary orbit is a limited natural resource. You can only put up certain things at certain times, and you have to know what's up there, and the scientists and engineers furnished a lot of this information.

KEN: Did you draft most of the treaties?

EILENE: No, I didn't draft them, but I wrote analytical reports on the provisions that had to be considered. I worked for the Space Committee and Foreign Relations Committee in the Senate when they were considering ratification. The treaty was ratified October 10, 1967. Lyndon Johnson was very powerful in this process, and this showed that he was a leader for our national policy, organization, program, and funding.

Working on Outer Space Treaty Issues

EILENE: The other treaties that I worked on are "Assistance and Return of Astronauts and Space Objects" and "Liability for Damage and Registration of Space Vehicles." The last one, the Moon agreement, which the U.S. and other spacefaring nations didn't ratify, was not successful. Nations comply with treaty provisions in order to preserve an orderly space environment. With space, you have more chance of success with an international agreement than with almost any other subject. You have about the same chance as with the postal service, because if you didn't observe regulation, there would be chaos, and the chaos would be in communications, not only in outer space, but on the Earth.

The process of making decisions by consensus is very important. I said to Ambassador Wyzner, who was the Chairman then, "I don't see how consensus is going to work." "Well," he said, "I'm a little worried about it too." The way it worked was that we had a text, and everything that any nation didn't agree with was put in square brackets. When the meeting was over, all the delegates had this to take back and work on until the next meeting. You could see what delegates agreed with and what they didn't agree with, and then each nation could work on the text in the months that intervened between sessions. I have written an article on this: "Consensus Decision-Making in the United Nations." A lot depends on the Chairman.

Now, Ambassador Wyzner didn't try to push anybody—he was very laid back. He was well aware of everything that was going on, but once we were in Geneva for a month because of the consensus process. After three weeks, Ambassador Wyzner said, in a very offhand way, "Now, we have been here for three weeks, putting in square brackets, and I'm in your hands, of course, I'll do whatever you want, but I was just wondering whether we could spend the last week removing some of the square brackets." And there we were in the springtime in Geneva with all the tulips and the swans on the lake and the watches and the chocolate bars *(laughing)*, and we couldn't go home without agreeing on something, so we got very busy the last week and were able to agree on deleting some of the square brackets.

There were some comical times when the U.S. and the Soviet Union would agree, and the delegate from India would get up and disagree about something, and then we had to go back again until we achieved a consensus. I learned a lot. You have to be very patient and you should never send anyone there who is impatient. We had one chairman of a subcommittee who had this gavel, which he liked to hit, and that annoyed everyone. Even if they agreed with him, they didn't want to favor his position *(laughter)*.

ROBBIE: What is the present situation?

EILENE: Some groups plan to review the treaties as to whether they need any revision. I am very anxious about this because there are some people who are purely legal, other people who are very commercially minded, and we need to strengthen the interdisciplinary process with space scientists and engineers. As long as you have treaties that take into consideration space science and technology, then you get along all right. People comply with them because they *have* to—just like you comply with the stop and go signals when you're driving a car. If you get people wanting laws for special philosophical ideas, difficulties arise. The trouble with the Moon agreement is that it provides that the Moon is the common heritage of all mankind. The term is not defined—it means different things to different people. It's connected with establishing some kind of international regime. So you get articles written by people who assume there will be a lot of commercial activity, that you can get on the Moon and make money, and apparently they think you can take a shovel up and dig in the Moon and get some diamonds and bring them back to Earth and sell them and distribute the proceeds equally throughout the world. They aren't specific about the definition of natural resources.

For example, I have this article on the commercialization of space: "Needed: A Private Property Standard for Space." They do not tell you what it is on the Moon or any celestial body that they are planning to mine. The word "mine" means different things to different people. The only thing that is really set forth that you can understand is solar power. Solar power doesn't go along with private property on the Moon. These people have got to tell me what they want on the Moon, and how are they going to do it, and how much is the cost, and how are we going to get there and bring back products to the Earth. (8)

KEN: And so today, this is open-ended, and not resolved.

EILENE: Yes. This is a good article in pointing out some of the differences between national and international, and some of the things that have to be done. But the psychology is based on the private property ideas that we have on Earth, and they don't necessarily translate into space

KEN: If you were to be in charge of the next session of the United Nations addressing this problem, how would you go about it?

EILENE: I would set up criteria, and get people to agree to the criteria. That is—are you going to amend the 1967 treaty, or are you going to make a protocol to it, or are you going to leave it the way it is, as general principles, and then when we want something specific, go ahead and make another treaty—that's what we have been doing. In making these other treaties, we have repeated the main provisions in the 1967 treaty, and therefore, we have a consistently developed body of space law.

One of my criteria would be that any document would go to the U. N. Scientific and Technical Subcommittee, with issues identified, before it gets to the U. N. Legal Subcommittee. I would not have it go to the Legal before it went to the Scientific and Technical. That's what we're supposed to do—that was the way it was set up. They didn't do that with the Moon. They had a number of developing countries coming up and saying, "Well, you have benefits for the spacefaring nations and we don't have special benefits." So that's the way—I would go about it slowly—I wouldn't be snowed by some of these people who have an axe to grind.

KEN: When you talk about the science and technology focus, which started a lot of this outer space initiative, they don't necessarily represent the commercial interests. Surely the commercial people would come in and have an input somewhere in this process. Or would they?

EILENE: Yes, I think the pharmaceutical companies would, for example. The pharmaceutical companies are among those who expect to benefit from the space station. They will have a lot of research there—that's purely scientific research. Then they can take the results and make money from it, so at first it's scientific, and then it's commercial.

ROBBIE: Can I go back to the time after Lyndon Johnson addressed the UN on creating the space committee? From 1968 on, can you give us a chronological sequence of what happened in your own personal career? Just because NASA was formed, you did not stop working in space. You expanded.

EILENE: Yes. I'm still in orbit! *(laughs)*

ROBBIE: You look so grounded to me. It's amazing how you can be both places at once! *(laughter)*

EILENE: My husband was in charge of the United States delegation of the members of Congress to the Interparliamentary Union, which is a meeting of all the members of all the Parliaments in the world. We took eleven such trips. I wrote the briefing book for the members.

ROBBIE: Your marriage and your Congressional research meshed rather nicely! After you quit working for the Congressional Research Service and the Senate, then what did you do?

EILENE: I'm Honorary Director of the International Institute of Space Law. I was Vice President of the Institute for a while, so I spent a lot of time on that, and also for the International Academy of Astronautics. There's a lot of work connected with it. I just really never retired—I've worked all the way through. For some years I was President of the Theodore Von Karman Memorial Foundation.

ROBBIE: What is your role in the International Institute of Space Law?

EILENE: Every year we have a colloquium on the occasion of the International Astronautical Federation Congress. The last one was in Turin, Italy, and the one before was in Beijing, China, and before that was in Oslo, Norway. There's a lot of work connected with this. In the case of the Academy, I was made Trustee Emeritus. I was in charge of Section

Four for nine years, relating space science and technology to the social sciences. That is my most significant contribution to interdisciplinary research.

Now I do a lot of work for the United Nations Office of Outer Space Affairs. In January, the Austrian government and the U. N. Outer Space Affairs Office invited me for a week in Alpbach, Austria—this was last January—to a workshop. They invited thirty people from different countries with different backgrounds. We worked on space futures and human security, and I wrote a paper on that subject.

ROBBIE: Can you say more about interdisciplinary research for space studies?

EILENE: Every space program involves a combination of disciplines—science and technology as well as the social sciences—economics, law, government, and politics, international relations, sociology, organization and management, etc. We may need a sociologist to evaluate relations of crew members in spaceflight, and a social psychologist to estimate the support of public opinion. When you are planning a space project, you need to identify every ingredient that is required for success and give each one a weight according to its influence on the results.

We seem to have two kinds of experts—specialists and generalists. My observation is that specialists think vertically within their discipline, while generalists think horizontally and can relate interactions between different parts of a program. We need both kinds of thinkers. When we are planning a space program, the unique controlling force of scientific "laws" must be complied with. You can adjust funding and human-made laws, but satellites can only produce information between the earth and outer space under known specific conditions. Space activities are dangerous, risky, and expensive, and need to be regulated, so the combination of elements needs generalist thinking.

ROBBIE: And clearly you are a generalist with familiarity with a number of different disciplines, which has been one of your great strengths in the space research you have done.

EILENE: When I look back, I realize that I did not have a plan for developing research about outer space, but I can now see a line of circumstances that led to this profession. In high school when I was Captain of my Debate Team, I was taught to study a political subject and learn every point of view and then discuss it pro and con. At Swarthmore College, the honors course in the social sciences was based on developing political science, economics, history, law, and philosophy—these were related fields to apply to any public problem. Then when I worked for the research arm of the Congress, I was in a perfect setting to do research connected with legislative actions, and they all had to be integrated and coordinated.

For example, hearings are held on authorizing a program, and then additional hearings are held by another committee on appropriations. When I was a National Defense Analyst, I had to analyze proposed legislation from the Department of Defense that affected the Army, Navy, and Air Force. The departments could be affected in different ways and had to be made equitable. When the main space treaty was sent to the Senate Foreign Relations Committee in 1967, I combined my work with the Space Committee. The committee print I wrote then was used by both Committees and our Ambassador to the UN when they were considering ratification. Space communications was of concern to a number of committees and presented opportunities for coordination of policies. I was fortunate to be in a workplace where overall horizontal thinking could be developed.

This carried over to my activities after I retired from the Congressional Research Service. I have already mentioned my work for the International Academy of Astronautics. As chairman of the Section on the Social Sciences, it was my responsibility to coordinate programs with the other three sections: Basic Sciences, Life Sciences, and Engineering Sciences. I was elected to this position for nine years. In my present appointment to NASA's Advisory Committee on the International Space Station, there are opportunities to relate national and international aspects of science and technology, the budget, and future planning.

KEN: Your experience is just absolutely phenomenal. Can you think of any advice or suggestions for those who might be at the forefront of space activities today?

EILENE: Space activities are developing along two main lines: exploration and uses. You could choose a career entirely devoted to outer space, or you might choose an activity that requires a space background to perform part of its function. In either case, you need to have a basic knowledge of space science and technology, because these facts can determine what you can and cannot do. Every scientific and engineering discipline is involved with space activities, as well as the social sciences and humanities. I wrote an article about how one discipline is involved: "Government in Action: The Role of Political Science in Outer Space Activities."

If you are a specialist, you need to know how your particular work fits into the whole field. If you are a generalist engaged in interdisciplinary research, you must be able to identify all the disciplines required for a project and use imagination in figuring out interactions and probable consequences.

ROBBIE: Is there anything else we should have asked you about that you might want to include?

EILENE: Yes! In 1982, I was invited by the Soviet Academy of Sciences as a guest when they celebrated the 25th anniversary of Gagarin's space flight. I was one of a small group from different countries, and we were accompanied by cosmonauts to Star City, where we saw their space training buildings. As we went about Moscow and its environs, I was impressed by the enthusiasm of the people for their satellites. It was almost as if their space program had become a religion.

I wrote a paper on "Conditions Essential for Maintaining Outer

Space for Peaceful Uses" which came to the attention of the U.N. University in Tokyo. The Vice Rector of their Global Learning Division, Edward Ploman, asked me to plan an arms control symposium on the theme of my paper. The symposium was sponsored by the International Institute of Space Law, the U.N University, and the Hague Carnegie Foundation. We invited about 30 experts from different countries, representing a variety of disciplines. We held the conference in the Peace Palace at The Hague in 1984. I wrote the first chapter on "Overview" and another one on "International Institutions to Ensure Peaceful Uses of Outer Space." The proceedings were published in a book—*Maintaining Outer Space for Peaceful Uses*. This was edited by N. Jasentuliyana, who is now Director of the United Nations Office of Outer Space Affairs in Vienna.

In 1994, the European Space Agency invited me to participate in their International Lunar Workshop, called "Toward a World Strategy for the Exploration and Utilization of our Natural Satellite." This conference was held in Beatenberg, Switzerland. I wrote the interdisciplinary discussion paper on "Political, Legal, and Economic Aspects of a Return to the Moon."

ROBBIE: What have you been doing most recently?

EILENE: The NASA History Office had a symposium on "Reconsidering Sputnik after 40 Years," and asked me to present a paper on "Organizing the U.S. Government for Outer Space: 1957–1958." Then I flew to Turin, Italy, where the International Astronautical Federation was holding its annual congress, and I gave a paper on "The United States and the 1967 Treaty on Outer Space." This happened last September and October [1996].

ROBBIE: What do they want you to talk about?

EILENE: Well, they got interested in my reaction about Sputnik—they want my analysis of the consequences of the Soviet Union being the first to orbit a satellite. I seem to be the only one who pointed out that we've had forty years of peace, as a result of the policy, the organization, the implementation, and the budget. Somehow they had taken that for granted! It was a remarkable achievement, and it was such a short period of time—between October of 1957 and the end of July, 1958, we had NASA up. By the end of 1958, we had a new Administrator of NASA, and it was full speed ahead. It was a *remarkable* period.

Space Debris

ROBBIE: Dr. Galloway, how long have you lived in this house?

EILENE: Since August of 1941. We moved here in August and I started working at the Library in September.

ROBBIE: So it all happened right here. No wonder you have so many documents stored here! What will you do with all your documents on space? Are you leaving them to the Library of Congress?

EILENE: No, mostly my son comes and paws through them and takes them. He teaches a course on space. I'd like to put some of my records on computer. I have all these papers in the garage, the basement, the study, the attic, that bedroom and part of this room. It's *all* on outer space. There will be a headline in the *Washington Post* some morning. It will just say "Old Lady Suffocated by Outer Space Debris," and that's the way I'm going to end!

[*Editors' note:* Eilene Galloway passed away in her home on May 4, 2009, at the age of 102. Her ceiling did not fall in, after all, and so she was not "suffocated by outer space debris"!]

REFERENCES FOR CHAPTER 4 AS PROVIDED BY EILENE GALLOWAY

1. *International Cooperation and Organization for Outer Space: Staff Reports* (Senate Document 56, 89th Congress, 1st session. 1965. 580 pages)

2. The full reference to the proceedings of these hearing is: Hearings before the House of Representatives Select Committee on Astronautics and Space Exploration HR11881 1958 85th Congress, Second Session on HR11881 April 15, 16, 17, 18, 21, 22, 23, 24, 25, 28, 29, 30, May 1, 5, 7, 8 and 12, 1958. Printed for the use of the Select Committee on Astronautics and Space Exploration, United States Government Printing Office, Washington, 1958. Eilene's report is on pages 5 to 10, titled "The Problems of Congress in Formulating Outer Space Legislation."

3. The full reference here is *Hearings, Reports, Laws and Committee Prints, 85th Congress, Second Session, 1958, Special Committee on Space and Astronautics, U.S. Senate.* The title page says, *Hearings before the Special Committee on Space and Astronautics, United Sates Senate, 85th Congress, Second Session on S.3609. A bill to provide for research into problems of flight within and outside the Earth's atmosphere, and for other purposes. May 6, 7, and 8, 1958. Part One.* Printed for the use of the Special Committee on Space and Astronautics, U.S. Government Printing Office, Washington, 1958. Lyndon Johnson is the Chairman of the Special Committee on Space and Astronautics; Eilene Galloway, Special Consultant. Her paper is on pages 381 to 386, and is titled "Space Law, a Symposium Prepared at the Request of the Honorable Lyndon B. Johnson: Chairman, Special Committee on Space and Astronautics, U.S. Senate, 85th Congress, Second Session, December 31, 1958. Preface by Eilene Galloway." Selected articles arranged chronologically: High Altitude Flight and National Sovereignty by John Kyle Cooper;, "Who Owns the Universe?" by Oscar Schecter; "Airline Space" by Ralph Hienrich, etc. Galloway's preface is page v through viii, and the whole section ends on page 572.

4. The Senate documents are Parts I, II, and III of the *Inquiry Into Satellite and Missile Programs, Hearings before the Preparedness Investigating Subcommittee of the Committee on Armed Services, United States Senate, 85th Congress, First and Second Sessions, November 25, 26, 27, December 13, 14, 16 and 17, 1957. January 10, 13, 15, 16, 17, 20, 21, 23, 1958.* Printed for the use of the Committee on Armed Services. Printed by the United States Government Printing Office, Washington DC, 1958.

5. *Inquiry into Satellite and Missile Programs. Hearings before the Preparedness Investigating Sub-Committee of the Committee on Armed Services, United States Senate, 85th Congress, first and second sessions. November 25–27, December 13–17, 1957 and January 10, 13, 15, 16, 17, 20, 21 and 23, 1958. Parts 1, 2, and 3.*

6. Senate Report Number 100 of the 86th Congress, first session. *Final Report of the Senate Special Committee on Space and Astronautics pursuant to S. Res. 256 of the 85th Congress, March 11, 1959.*

7. Number 44, *United States International Space Programs: Texts of Executive Agreements, Memoranda of Understanding, and Other International Arrangements 1959–1965.*

8. "Needed: A Private Property Standard for Space—Continuing the Property Rights Debate" by Lawrence T. Roberts, in *Ad Astra*, Volume 9, Number 4, November–December, 1997.,

CHAPTER 5

PAUL G. DEMBLING
LAWMAN OF SPACE POSSIBILITIES

BIOGRAPHIC INFORMATION

PAUL G. DEMBLING (1920–2011) was a retired partner of the law firm Schnader, Harrison, Segal & Lewis. He received his AB degree Cum Laude and with Special Honors in Economics, and his MA degree from Rutgers University, where he had served as graduate assistant and teaching fellow. He obtained his Juris Doctor from The George Washington University Law School, serving as an editor of the *Law Review*. He was a member of the District of Columbia Bar, U.S. Court of Appeals for the District of Columbia, U.S. District Court for the District of Columbia, and the Bar of the Supreme Court of the United States. When Mr. Dembling entered the Federal service, he initially occupied various industrial relations positions. Later, from 1951–1958, he served successively as Special Counsel, Legal Advisor, and General Counsel for the National Advisory Committee for Aeronautics (NACA). In the latter capacity, he was the principal drafter of a bill which became the National Aeronautics and Space Act of 1958.

Upon the formation of NASA, Mr. Dembling was appointed the Assistant General Counsel, and served in that position until 1961.

In addition, he was the first Chairman of the NASA Board of Contract Appeals (1958–1961). He was subsequently designated Assistant Administrator (Director of the Office of Legislative Affairs) 1961–1963, during which period he continued to serve as Vice-Chairman of the NASA Inventions and Contributions Board (1959–1961). (This statutory Board determines the property rights of contractors in any invention made in the performance of any work under a contract with NASA, and grants licenses for any invention for which NASA holds a patent. The Board also makes monetary awards for any scientific or technical contribution to NASA having significant value in the conduct of aeronautical and space activities.) From 1963 until 1967, he served as Deputy General Counsel. In 1967 he was named General Counsel. He remained in that post until 1969, when he was appointed Deputy Associate Administrator of NASA.

From 1964 until 1969, Mr. Dembling also served as a member and alternate representative of the United States delegation to the Legal Subcommittee of the United Nations Committee on the Peaceful Uses of Outer Space in the drafting of the *Outer Space Treaty of 1967*, the *Treaty on Rescue and Return of Astronauts and Space Objects*, and the Convention on International Liability for Damage Caused by Space Objects. In November 1969, he was appointed General Counsel of the General Accounting Office (Office of the Comptroller General of the United States) and served in that capacity until 1978, when he entered private practice to head the Washington office of the Schnader, Harrison, Segal & Lewis law firm.

Mr. Dembling was the recipient of the Meritorious Civilian Service Award of the War Department (1945); and NASA's highest award, the Distinguished Service Medal (1968). He was presented the National Civil Service League Award in 1973, and was elected to the National Academy of Public Administration in that year; he then became a Fellow of that Academy. In 1982, the George Washington University Law Association presented Mr. Dembling with its Professional Achievement Award. In 1992, he received the International Institute of Space Law of the International Aeronautical Federation Award "in recognition of his distinguished contributions to national and international space law."

Mr. Dembling was also a member of the American Bar Association. He served on its Council of Public Contract Law 1983–1984, was Vice-Chairman of that Council from 1984–1985, Chairman Elect from 1985–1986, and Chairman 1986–1987. He was a former chairman of the Model Procurement Code Coordinating Committee; a Life Fellow of the American Bar Foundation; Chairman, Public Contract Law Committee, Administrative Law Section 1979–1981; Chairman, Federal Acquisition Regulations Coordinating Council 1981–1983. He also served as a member of the following:

- International Institute of Space Law (President of the U.S. Association, 1971–1972);
- National Contract Management Association (Fellow; member of the Board of Advisors, 1971-present).
- National Contract Management Foundation (Trustee 1975–1980; Treasurer 1981–1992);
- Bureau of National Affairs, Federal Contracts Report Advisory Committee (1978–1995);
- Federal Bar Association (Editor-in-Chief, The *Federal Bar Journal* 1962–1969; member, National Council, 1964–present; President, Capitol Hill Chapter, 1977–1978; National Secretary 1978–1979; Vice-President 1980–1981; President-Elect 1981–1982; and National President, 1982–1983.)
- D.C. Bar (member, Steering Committee, Government Contracts and Litigation Section, 1989–1995; Procurement Round Table Board of Directors 1984–1988; Vice-Chairman 1988–1995).

Mr. Dembling served as Professor and Lecturer at George Washington University National Law Center for 22 years, and later served as Lecturer on Government Contract Law for American Graduate University/Procurement Associates; he had previously lectured on Government Contract Claims and Debarment and Suspension of Government Contractors, Federal Publications, Inc. He was co-author of *Federal Contract Management 1988* and *Essentials of Grant Law Practice* (1991), and contributed many articles to professional journals.

INTERVIEW

This three-hour-long taped interview, conducted on Nov. 21, 1997 in Dembling's office in Washington DC, was edited by both interviewers and by Paul Dembling himself for clarity, organization, and flow. Paul also provided some written additions, which have been incorporated into the text or appear in the full text of the interview, which can be found at www.davis-floyd.com.

Childhood and College Years

ROBBIE: Let's begin with your early childhood, to see if there was anything that made you interested in outer space.

PAUL: I was born and raised in New Jersey. I went to the elementary schools and to Rahway's only high school. My father was in the textile business and my mother was a homemaker. They met and married in New York. They were both working there and moved to New Jersey because they had friends there and had decided to raise a family away from New York City.

ROBBIE: What were you into when you were a kid? What did you think and care about?

PAUL: At one time, I was thinking of becoming a physician. In my high school yearbook it says, "Dembling will become a purveyor of pills" or something like that. I went to Rutgers and majored in Economics and did pretty well. This was during the Great Depression. I found by that time that I was not really interested in medicine. I was going to major in Business Administration and then I switched over to Economics, primarily because of the teaching staff in that department. There were some outstanding people in Economics at Rutgers at the time. Arthur F. Burns was one. He later became Chairman of the Council of Economic Council Advisors to the President, Chairman of the Federal Reserve Board, and served as Ambassador to Germany. He taught at Columbia as well as at Rutgers. He was also serving as Deputy Head of the National Bureau of Economic Research under Wesley Clare Mitchell and later succeeded Mitchell as head of the Bureau. Several other people who were later involved in economic policy in the federal government also taught economics at Rutgers at that time.

Burns was my mentor. He gauged what I was doing and said, "If you go on and get your Master's and PhD, we'll find a spot for you at the Bureau [meaning the National Bureau of Economic Research] so you can work there, and we'll make sure that you start with an Assistant Professorship at the school." That was my goal. I got my Bachelor's. Then, while holding a Teaching Assistantship, I got my Master's. The teaching assistantship was a little unusual, because it was only half-sponsored by the Economics Department. I'd done some work for Professor Marden, the head of the Sociology Department, so he wanted to give me a teaching assistantship also. I wanted both but the school didn't want to pay for two assistantships, so they said, "The way we'll solve it is you can work half time in the Sociology Department and half time in the Economics Department," so that's what I did. What that turned out to mean was that I taught a course in sociology and marked papers in the Economics Department.

I graduated from Rutgers in 1940 and liked the academics so much I just stayed on. One of the other schools that offered me an assistantship was Clark University, where Robert Goddard worked. I don't think I knew of Goddard at that time, but I later learned that had I gone to Clark, I might have run into him then.

In the War Department 1942–1946

PAUL: By the time I received my Master's in 1942, the war was on.

I remember Pearl Harbor Day, December 7, 1941, very well. I was home marking papers that day and my folks had gone to the movies. They went to see *Easter Parade* with Judy Garland. It's strange how you remember things like that—the events of the day imprinted that permanently.

I managed to stay at Rutgers from December of 1941 until May of 1942 when I got my Master's, and then all bets were off. Since Rutgers was a men's school at the time, the draft had a big impact—more students were being called up every day.

I started looking around to see what was going to happen. I assumed I was going to be drafted, and in June of 1942, I was called up and then deferred because of a bad knee. I had had a split kneecap, an old football injury, and during the physical exam they gave me a fifty-pound pack and had me pick it up and walk around with it. They looked at my kneecap and saw that it separated, and they said, "Well, he's going to have problems and we're going to have to be responsible for him. So don't induct him in the first place!"

Sometime previous to that, I'd taken the Federal Government's Junior Professional Assistant examination and had done well enough to be offered several positions, so I came down to Washington and accepted a job with the War Department in the Transportation Corps. Several weeks after I entered on duty, my draft board in Rahway, NJ sent me another notice. So I took the physical again and they found that the knee hadn't improved. So, at the conclusion of the exam the doctors said, "The Army can't use you," and I returned to Washington.

ROBBIE: Were you relieved? Did you want to be a soldier?

PAUL: I don't know. At that time, everybody was going in so I felt left out. We were "4Fs," if you remember that term. It was sort of a blemish, and you felt as if you weren't quite good enough. As I said, I had several offers from the Federal Government and when one from the War Department came through I accepted it.

ROBBIE: Did you feel this way you could make a contribution to the war effort?

PAUL: Yes, that was the idea. In fact, I got a job in the Office of the Chief of Transportation Control Division as a Statistical Analyst because of my economics background, and because of my rating in the Junior Professional Assistant Examination (Economics). We compiled statistics regarding troop locations, troop movements, and the location of troop and cargo ships. Actually, during the war, the Army had more ships than the Navy, and we kept track of those. The Joint Chiefs of Staff used the information, and our monthly report went to about twenty people in the government, including the President. The material was so sensitive and confidential they had us working in locked vaults. The colonel who used to deliver the reports had them chained to his wrist and they'd unlock them at the receiving end.

KEN: It was that high-level?

PAUL: Yes. There was a second lieutenant whose only job was to maintain code names for all the places where we were fighting or had targeted. He maintained this code book and changed the codes periodically. Often, we weren't even sure what the place was. You know, Hawaii might be "Batter Up!" and Christmas Island might be "Strike Out" or some other equally imaginative code term. They didn't want the code names to be anything that could be associated with something else—in other words, you couldn't use all baseball terms to refer to a particular set of places or planned events, because somebody might figure that out. The Normandy Invasion, for example, was initially called "Bolero."

Every month we put together a book containing all kinds of information relating to the Army and relevant to the war, using graphs and bar charts to show the data—oil levels, fuel levels, the location of casual soldiers (those not assigned to a regular unit), replacements, troop supports, and the location of cargo ships, inter-island ships, and troop ships. Telegrams would come in from all over saying, for example, "45 troops came in to replace those that were killed last month." We compiled all the information showing our military strength all over the world.

Often information wasn't made public until much later. I remember that the President Coolidge, which was being used as a troop ship, was torpedoed in October and the newspaper didn't carry the story until the following May, and then it was written up as if it had just happened—"The President Coolidge was torpedoed." The story didn't say, "last night," or "several months ago." It was written in such an ambiguous way that someone reading it might be led to believe it had just happened.

ROBBIE: So you were making a *huge* contribution to the war effort?

PAUL: Well, I hoped so. I devised a method whereby the graphic artists didn't have to constantly redo their work. Every month they'd completely redo all their charts—whether they were graph, or bar, whatever they were, they would redo them. I said it was silly to redo the whole thing just because they sometimes changed the figures in the back. I said, "Why don't we make up a master one, then just add to that as we get additional information" —then even if you had to go back you could change it easily. They used that zip tone material and you could cut it and do that.

So, that's what we did, but it meant that I had more free time during the week. My colonel chose to spend his free time on the telephone. He'd been in the oil and gas business, so he was buying and selling stocks as I recall *(laughter).* I spent mine reading quarter pocket books. Since reading unofficial material in the office was unseemly, I had to read in the men's room! *(laughter)*

Those were interesting years. I did that until I realized that I was under-utilized. We'd gotten the system set up, and it was getting to be so automatic that I asked to be allowed to transfer out. I said, "Besides, you aren't really using my skills." The War Manpower Commission was recruiting, and I was interested because I'd done my Master's thesis on "Minimum Wage Legislation and Administration." It also offered a healthy increase in pay, but you couldn't transfer without approval from the releasing agency. There was a Skills Utilization Committee that decided where you would be more valuable to the war effort. They wouldn't release me for that job, but they said, "There's a job within the War Department we *will* release you for—in Wage Administration and Job Classification. Someone had examined my Master's thesis and seen the connection.

Civilians manned the troop transports, which were operated by the Army Transportation Corps, and this new job was interesting because it was in the division that negotiated contracts with the labor unions. I attended meetings with representatives from the maritime union and the longshoremen's unions on both the East and West Coasts—people like Harry Bridges and Joseph Curran. We also did wage surveys to determine how much people should be paid for jobs in the Marine Repair Centers. All the ports of embarkation that were under the control of the Transportation Corps had Marine Repair Centers.

When the ships came in, civilians repaired them for the most part, so we set up civilian pay rates for those people, and for the civilian seamen. There was also a system of bonuses for them—for example, there was a bonus for being on a ship that was in combat, and additional bonuses if the ship was hit, or if the seamen were injured. We also had to operate within the statutory price and wage controls in effect at the time, so we had to determine what the law allowed as against what was reasonable and justifiable under the wage surveys we conducted.

Robbie: How long did you stay in that position?

Paul: Until the end of the war. Then, the question was, "What next?" The people I'd worked with in the Control Division were officers who'd been commissioned directly out of civilian life. They'd worked for oil companies and steamship companies. Many of them advised me to go to law school, saying, "Young man, you know what you're doing and you do it well, but you don't know the problems we have in terms of negotiating and drafting an agreement. You ought to take a law course or at least a contract course to understand what we're going through."

My good friend from Rutgers, Martin Friedman, also suggested that I take some law courses. (Incidentally, he later became one of the three assistants to President Truman, along with Clark Clifford and Donald Dawson.) So, acting on his suggestion, along with that of my former bosses, I decided to start by taking a course in contract law at the George Washington University Law School. Once I started, I said, "This is for me," and went right on through.

The Post-War Years at NACA

PAUL: After the war, I transferred over to the NACA—the National Advisory Committee for Aeronautics—the organization that performed the basic aeronautical research for the country.

A lot of people in the Transportation Corps were going back to their pre-war jobs and I felt it was time to leave, so I started to look around. Martin Friedman told me he had served with Robert Lacklen, who had just been named Director of Personnel at NACA. NACA's headquarters was here in Washington. I interviewed with Bob Lacklen and said, "Here are my credentials," and he said, "Great, I'm forming an office and a team—come on over." So I joined him.

NACA Headquarters consisted of about 125 people. All the operations had been carried out in the field laboratories—Langley Aeronautical Laboratory in Virginia, Ames Aeronautical Laboratory in California, and Lewis Flight Propulsion Laboratory in Ohio. There were 8,000 people working in those labs and they were apprehensive about us because we weren't scientists. They said, "The only way you're really going to understand what we do down here is for you to spend some time in one of the laboratories. The place where you're going to learn the most about us is in the Structures Division because it's the most theoretical." So a decision was made for us to spend a month at Langley.

When we were pushing for NACA to become the foundation for the space agency, we checked on how much time and effort was being spent on outer space research, because that was one of our justifications. We found that NACA was spending about 25 percent of its time and money on space back then.

KEN: Was that a surprise?

PAUL: It wasn't a surprise to the directors and to people who knew. It was a surprise to those who were not intimately involved in the scientific aspects. The administrative staff, as opposed to the engineers and scientists, didn't have to know.

Within NACA, I set up job evaluation and classification programs. We set standards—for example, we coined the title of "Aeronautical Research Scientist." Previously, all titles were taken from the disciplines in which the individual had studied—"Physicist," "Civil Engineer," or "Mechanical Engineer." We defined what such a scientist did: what should an Aeronautical Research Scientist get paid? How do you know the difference between a first-level scientist and a journeyman? We created a civil service examination for this category and that made recruitment easier. Then other agencies copied what we had done—the National Bureau of Standards, the Navy Department, various others. Many of the jobs were thinking jobs. The people didn't build anything; they didn't show anything for what they were thinking about. So we had to look at and define both what they were thinking about and what was unusual about one person's thinking as compared with another's. Some of the young people joining the organization were solving projects that had been around for a long time and had gone unsolved by older heads.

ROBBIE: Can you give me an example?

PAUL: Yes. A fellow by the name of Bernie Budiansky, who later went on to Harvard, solved a couple of problems that the top scientists at Langley had not been able to solve. There were other examples of young people that were very good, but he's the one I remember best. In 1946, when I came over to Industrial Relations, we did those kinds of things. We worked out a little formula for ourselves: *you could make an assumption that was unusual and not correct and it wouldn't help much, but people who made assumptions that were unusual, innovative, and correct were the ones who developed ideas and made the most significant contributions.*

It was also very interesting that, for the most part, they made their name or their early contributions before the age of 27. If they hadn't really made a big hit by 27, they were going to be OK, they could continue to make contributions, but they weren't going to make that big discovery. Even Einstein made his first big discovery before he was 27.

KEN: In some ways, the Langley environment was very unusual in that sense. Is that a fair way to put it?

PAUL: That's right, but that was true of all the NACA laboratories. When people went to school to become aeronautical engineers, their first choice was to go to work for NACA because that's where they were going to learn their profession, their craft. I heard from most of the people on the staff then that when they were at engineering school, the generally agreed-on goal was to obtain a job at NACA. It made for the start of a successful career, whether they stayed in government or went to industry.

ROBBIE: I've always heard of Langley Field in those days as being the gathering ground for brilliant, creative minds who flocked there because that was where the action was.

PAUL: That's right! NACA had its choice of the best students, and it took them.

Law School

PAUL: After the war, I attended law school at night. My routine was to work at NACA until about 5:30. Classes started at ten of six. I would go into the men's room and wash myself with cold water, just to freshen up. Then I'd walk across the street, get a cup of coffee, and go on to school for two hours, and then I would go home. By that time I was married, so my wife would serve dinner at about 8:30 or 8:45. We might talk or I might look at the paper. By 9:30 or 10:00, I'd sit down and try to get two hours of studying in so I could keep up with the classes.

ROBBIE: This was after two hours of classes and a whole day of work.

PAUL: Right! I couldn't do it today. I did that every night of the week except Friday. Otherwise I wouldn't have been able to catch up in law school. Friday night was free time, and we might go to a movie, but Saturday and Sunday I'd have to put in about three hours each day. I didn't want to be like some of the people I'd seen over there going to school five, six, eight, ten years. I made up my mind I was going to go right through. I went summers and winters, and in two years and eight months I'd completed the program. It takes three years if you go to day school, and if you go to night school it generally takes at least four years. I was Labor Editor of the *Law Review*. I also wrote two articles that were published during that time.

ROBBIE: Did you have kids by then?

PAUL: No. I graduated in May of 1951, and my boy was born then. In fact, I dropped my wife off at the hospital and went to take my last exam.

ROBBIE: Did you miss the birth?

PAUL: What happened was that I went back to the hospital after the exam and the doctor said, "You're not going to do anything here. She's not going to deliver before the morning. Why don't you go on home?" So I went home and she delivered at 8:00 that morning.

ROBBIE: So you got your law degree, and you stayed with the NACA.

Were you still doing Wage Administration?

PAUL: Yes, by that time I was a Special Assistant to the Number Two man at NACA, the Executive Secretary, Dr. John F. Victory. The operating head of NACA was the Director, Dr. Hugh L. Dryden. Victory was head of Management and Administration, and then there were Associate Directors—one for propulsion, and one for structures and one for aeronautics.

The top policy and governing body was the Committee—similar to the Board of Directors of a corporation—consisting of representatives from the aircraft manufacturing and airline industries, the Navy, the Air Force, the Smithsonian Institution, and academia.

Graduation and Legal Work for the NACA

PAUL: I passed the bar exam and was admitted in 1952 to the District of Columbia Bar and to the Federal District and to the Federal Court of Appeals for the District of Columbia. I was the Special Assistant, but they started handing me some legal problems.

One day, I was told that the General Counsel had resigned, and they asked if I'd like to move into his slot. They said, "You're better off if you're not called "General Counsel"—why don't we call you "Legal Advisor," like they do over at the State Department." So that's what I was, and then, later on, I was called General Counsel.

ROBBIE: So you became the Legal Advisor for all of the NACA!

PAUL: Yes, then later on I was named General Counsel, and did all the legal work.

ROBBIE: What were some of the issues that you had to tackle in the early 1950s?

PAUL: I did a lot of work with Congress—legislative matters that affected or could affect NACA. Contractor disputes represented another area of concern. In fact, we established a Board of Contract Appeals, which I chaired. I worked with the people in the Agency on drafting their testimony. I was the liaison with the Appropriations Committee of the House and the Senate. That stood me in good stead, because when we became NASA, and when Jim Webb came on board as Administrator, he was looking for someone to handle Congressional relations.

Webb didn't want a political person in the job of Director of Congressional Relations (it was called Legislative Affairs) so he consulted with Dr. Dryden, who was, by then, Deputy Administrator of NASA. He asked John Johnson, the General Counsel, and he asked Jim Gleason, the man who was leaving the position. Webb said he was looking for a person in the agency who knew the Hill, who'd worked with the Hill, who understood the Agency, and would be a good person to put into the job. He later told me that my name showed up on all three lists. He didn't know me, and I didn't know him. He just called me in and said, "I'd like you to be the Director of Legislative Affairs—you've been recommended by three top people in NASA."

ROBBIE: Was that a higher position than General Counsel?

PAUL: I wasn't NASA General Counsel at that time. I was the Deputy General Counsel. I told Mr. Webb, "I'm happy doing my legal work, being Deputy General Counsel," because I thought I was going to lose the opportunity to become General Counsel. And his reply was, "Well, why don't you do this job until we reach a plateau—you get it organized, and then you can go back to 'lawyering'"—that's what he called it. So, that's what happened. I was in the Congressional job for close to three years, from 1961 to around 1964.

Sputnik and the Beginning of the Space Program

ROBBIE: What position were you in when Sputnik happened?

PAUL: I was General Counsel of NACA. Sputnik went up in October of 1957, and naturally it caused quite a stir in the scientific community. The Navy Department had been scheduled to launch its own satellite, the Vanguard, during the International Geophysical Year, but hadn't yet done so, and there was great concern that the Russians had beaten us to it.

ROBBIE: Did anybody have a clue that the Russians were working on Sputnik? Was it just totally out of the blue?

PAUL: I think what was out of the blue was that they were that close to launching. They were part of the International Geophysical Year, and I'm sure that everybody knew that every nation of the IGY was working on something. I don't think anyone knew how fast they were working or how close they were to launching. It went up and it worked, and that raised the question of what was going to happen in this country. How were we going to react? Eilene Galloway probably told you about the hearings that were held under Senator Johnson and his Disarmament Committee. There were hearings on the Hill, immediately—in the House, and in the Senate, regarding Sputnik. In the Executive Branch, there was a question as to what the President would do.

KEN: Eisenhower's policy?

PAUL: Yes, initially, he wasn't very excited about it.

ROBBIE: He wasn't excited about Sputnik?

PAUL: He didn't think it was that big a deal. However, he did appoint a Science Advisor—he was the first president to have one—Dr. Killian, from MIT. Killian's job was to pull together a space policy and he came up with a concise and complete statement that spelled out the policy that Eisenhower adopted.

KEN: Did NACA provide input into that?

PAUL: Yes. NACA personnel provided input.

ROBBIE: What did Killian's policy say, essentially?

PAUL: Essentially, it said that there ought to be a new separate agency, and that it ought to be a civilian agency. This was very interesting, especially from a President who had spent his entire career in military. The policy further recommended that the function should not be added to any existing agency, i.e., the Department of Defense. That was the big decision, which I guess surprised a lot of people. Here was a military person saying, "I think it ought to be a civilian agency." Once the decision was made, of course, it didn't satisfy the Air Force, so they lobbied against the idea of a civilian agency. They thought they should have it. And the Department of Defense, as a whole, felt that it belonged there, that they knew how to handle this. They considered it primarily a military matter. Sputnik was sent up with a military booster, after all.

KEN: So it was a derivative of the ballistic missile.

PAUL: That's correct. One reason we had a problem launching some of the early spacecraft was that we had gone into much more powerful, smaller and more efficient boosters for our weapons systems. The Russians were not yet technologically advanced enough to do it through smaller boosters in the weapons arena. They had the big boosters, which you need to put these spacecraft into orbit. The fact that we were so much ahead, technologically, actually hurt us in terms of not having a big enough booster to lift the spacecraft.

ROBBIE: So the reason Sputnik was so big had more to do with it being a cruder kind of thing, but it made everybody think that we were behind and inferior.

Paul: That's right.

Robbie: Maybe we should have officially thanked the Russians for galvanizing the American space effort!

Paul: I recall seeing Dr. Dryden and Dr. Sedov, his Russian counterpart, talking once about how much they were each doing, and how much each had to do and so forth, and the Russian saying, "Dr. Dryden, I think that we help each other out much, by your moving ahead and then we move ahead, and we both get money out of our legislatures!" *(laughter)*

When Sputnik went up, the NACA people were upset, especially the ones known as the Young Turks. They thought we ought to *do* something. NACA was already spending 25 percent of its time and money on space research, so why shouldn't we be doing it? Clearly, we were the ones who ought to be the chosen organization for the space program.

They went to Dr. Dryden and said they wanted a meeting with Dryden and Dr. Doolittle, the Chairman. As I said, NACA was organized like a corporation. There was a Board of Directors that ran the organization, and below that was a Director who functioned like the President of the Corporation. Dr. James ("Jimmy") Doolittle was the Chairman of the National Advisory Committee.

Now, Jimmy Doolittle was always known as a "flyboy." He had won the airplane races in Cleveland, and was the first one to fly across the country and set a world's record. He was the bomber of Tokyo during World War II, and the head of the 8th Air Force. People don't realize, though, that he had an earned Doctor of Science degree from MIT. He was also the one who encouraged Royal Dutch Shell to go into the high-octane business to supply aircraft with fuel early on. Most companies felt that airplanes were a sportsman's business, a sportsman's toy, but he saw the value of airplanes and that's what made him rich, and the company rich as well. So when he was Chairman of the NACA, he was not just a flyboy, he was a Doctor of Science.

Robbie: He was a very astute businessman as well.

PAUL: That's right. He and Dr. Dryden agreed to hold a dinner meeting at a hotel where they invited all those Young Turks who had been speaking out to come and express their ideas and their thoughts. This dinner took place on December 18th. All the Young Turks came and raised hell, agitating for action. I was there. There was an internal protest, which I would characterize as agitation:" Do something, *do something*!" They didn't want to be left in the backwash of the Sputnik tide. They wanted leadership. They felt that Doolittle and Dryden should aggressively pursue the chance for the NACA to become the space agency of choice.

The Space Act: NACA Becomes NASA

PAUL: Regarding Sputnik and the idea of a new agency, I had my own suggestion. I went to Dryden and said, "This town works on a draft that somebody prepares, so let me take a couple of weeks to prepare a draft of legislation that I think we might use to foster our own position."

KEN: So you came up with this as a suggestion—no one told you to do this?

PAUL: No one told me to do it. I volunteered and I asked Dryden whether I could do it. He said, "Sure, why not?" So I set about drafting the legislation, which was something that I felt I could contribute. I took an unusual approach to drafting a piece of legislation. For the authorities section of the bill, I reviewed every General Accounting Office (GAO) decision regarding agency authority. You may know that the GAO is an arm of the Congress with responsibility for interpreting the laws of the Congress regarding appropriations. This is the "watchdog" for the Congress, issuing decisions regarding the fiscal laws of Congress. The Comptroller General of the United States is the Head of the GAO, and decisions issued by the Comptroller General are binding. Agencies ask for decisions on whether or not their legislation permits them to undertake or perform a function, and GAO responds as to whether or not the authority exists in the legislation to do so and expend money for that function.

I reviewed all of those case decisions and placed those authorities in my draft of legislation, with the result that NASA ended up with more authority than any other department or agency that existed. I checked every authority that was listed, and where GAO said, "No, you don't have the authority because you don't have the language," I wrote down the language they said you should have if you want that authority. Of course, in cases where the decision recognized the statute as providing authority, that was also incorporated into the draft!

As a result, the Act covers *everything*. Every item in the NASA Act is covered by a decision of the General Accounting Office. I also looked at the Attorney General's decisions, though the Attorney General really doesn't render very many decisions regarding the finances of an agency or the financial authority of agencies. The GAO has that responsibility and the decisions are binding. A GAO decision can result in requesting the Treasury to withhold funds for a program, project, or whatever.

ROBBIE: Did you go to the GAO archives for all of this?

PAUL: They were all in the published decisions. I only used the decisions that were available.

They're collected in volumes that are published each year. The Comptroller General's decisions are available to the public in books printed by the Government Printing Office. They serve as precedents. If I had problems, I would call over to the GAO's Reference Service. Practically nothing was changed in the list of authorities I incorporated into my draft, though there was one change. GAO had rendered a decision to the National Bureau of Standards involving the acceptance of gifts. Since an agency's appropriations may not be augmented by acceptance of gifts unless so specified in its legislation, GAO had held that the Bureau had the necessary language to accept gifts. I had used the same language. When we were up testifying before the Senate Space Committee, Senator Clinton Anderson saw that and said, "Does that mean that General Electric could give a gift to NASA and say, 'This gift is for Dembling?'" *(laughter)*

I hadn't thought of that, and I told him I hadn't, so he said, "I want the word 'unconditional' in there, as in 'unconditional gifts.'" So that's the language that's in the law now. It says that NASA can accept only unconditional gifts.

KEN: There you go!

PAUL: Later, after Alan Shepard's flight, people started asking, "Why are we spending money in space? Why don't we spend it on housing, why don't we spend it on something else?" Dr. Dryden made a speech in which he said, "Space research only costs each person in the country a dollar and a quarter." We started getting checks for $1.25 from people all over, saying they wanted to pay their share. Later, during the Apollo program, the same thing happened. People would send in money, saying, "I want this used for the Apollo program," but we couldn't accept it.

KEN: Because that was not unconditional.

PAUL: That's right, it was conditional. Of course we wanted to keep the money—we didn't want to send it back! So we wrote them a letter that said, "We are not authorized to accept a conditional gift. We will return the gift to you if you insist that it be used only for the Apollo program. We can, however, accept unconditional gifts, and we assume that you would want it to be an unconditional gift to the Space Program, recognizing that Apollo comprises sixty percent of our program," or something like that. In other words, "If you want your money back, let us know!" and they didn't.

With a government agency, any monies you get from user fees or from programs you run for the public are called "miscellaneous receipts." You're required to turn them over to the Treasury Department. The reason, as I stated before, is that an agency cannot augment its Congressional appropriations—if you could augment the appropriations, you could influence the actions of the agency. NASA, however, had authority to keep that money if the work were performed under a Cooperative Agreement.

ROBBIE: Oh, that's perfect!

PAUL: That's right! The language in the NASA act states, "In the performance of its functions, the Administration is authorized to enter and perform such contracts, grants, leases, cooperative agreements or other transactions as may be necessary in the conduct of its work and on such terms as it may deem appropriate." Now, if you look at the Act, it grants the *Administration* these authorities. It doesn't say "the Administrator," because at that time there was a lot of controversy as to whether a Head of an agency could delegate certain authorities and certain activities. I thought, rather than becoming involved in such disputes, it would be better to let the Congress provide these authorities to the agency itself. This avoids any question as to the delegation of authority within the agencies. Furthermore, the Administrator can always pull it up to himself, and if it's vested in the Administration, anybody could carry out a function. So, that language was deliberate and I was quite pleased that it was enacted.

My son, Doug, attended a lecture a few years ago, and he came home and said, "Did you make a mistake by not putting in that the Administrator had the authority?" I said, "No, it was on purpose!" He was told that in every other agency the Secretary or the Head of the agency had the authority to delegate. I said, "Yes, but there were problems with how much he could delegate." The problem arises from what are considered inherent duties of a Head of an agency—he or she can't delegate certain functions. Therefore, there was always controversy over what he or she could or could not do. Could an Acting Administrator do it, or could a designee do it, or could a Deputy do it? On what authority? So I said, "Let's just avoid all that! Let's give it to the Administration, rather than to an Administrator or a Head of the agency." So, it says that all of this authority is in the Administration—you don't have to worry about who has it.

ROBBIE: What course was your son taking?

PAUL: It was a seminar in legislation.

ROBBIE: Were they were looking at the wording of the Space Act?

PAUL: Yes. This was a legislative example, an organic statute of an agency they were using.

ROBBIE: I bet he was thrilled he got to go back to his class and tell them why. What a treat! Okay, so tell us who came up with the name NASA?

PAUL: We wanted to keep it as close to NACA as we could. It was important to keep "aeronautics" in there because that was the base of the organization. The aeronautics people felt they'd been left out, with space getting all the attention. The draft bill referred to the organization as "the National Aeronautics and Space Agency," but House Majority Leader McCormick thought that "Administration" elevated the organization. At any rate, I think that calling it "Administration" was McCormick's idea.

ROBBIE: Actually we just found out whose idea it really was! It was Eilene Galloway.

She just told us yesterday. She said to McCormick that an "agency" wasn't important enough, and he said, "But we're already calling it NASA, so we have to keep the A," and she said, "Well, call it an Administration instead of an Agency." She wanted to elevate it and give it that extra cachet.

PAUL: In any event, there was a tremendous fight going on in this town regarding who would be picked to carry out this function. The Atomic Energy Commission was pushing very hard. The Air Force was lobbying very hard for it, despite the fact that Eisenhower had said that it was going to be a civilian agency.

I would go up to the Hill to help educate the people there regarding NACA's role and why it should be NACA that became the civilian space agency. Senator Styles Bridges was the ranking Republican member of the Senate Armed Services Committee. I wanted to see him, and he said, "I can't see you." I said "Why not?" He said "I don't want military personnel to see me talking to you."

KEN: Representing NACA. *(laughter)*

PAUL: He said, "But I'd be happy to meet with you." I found out that Victor Immanuel, who was head of a major defense company, would take a suite at the Carlton Hotel whenever he visited Washington. He was a very good friend of Senator Styles Bridges, so every time I needed to see Bridges, to bring him up to date on what was happening, we'd call Mr. Immanuel and he'd set up the meeting at the Carlton Hotel. Bridges didn't want the Air Force or the military representatives to see him visiting us at NACA, or me visiting him at his Senate office. The legislative liaison people were all over the Hill and he didn't want them to see us meeting with him. He wanted to appear to be neutral. He didn't know where things were going to come out, but he didn't want to be accused of supporting a civilian agency or even the concept of a civilian program. I don't know how he felt, but we were briefing him.

ROBBIE: So why did he want to be in touch with you at all? Why would he want to be briefed?

PAUL: I think he just wanted to know what was going on. We were briefing all the other people up there. Also, he would tell us some of the concerns that the military had. I like what Walter McDougall, in his book, *The Heavens and the Earth*, says about us during that period *[reading]:* "By mid-January, NACA Director Hugh L. Dryden, Doolittle and Chief Counsel Paul Dembling had in hand a coherent space program based on NACA in cooperation with DOD, NSF, NAS, universities, and industry. David challenged the Goliaths for the limitless and potentially richest fiefdom of them all—outer space."

KEN: Certainly it would have been of benefit to know about the military's concerns.

PAUL: Yes. When the bill was finally introduced, there were changes with regard to some of the prologues and some of the substance, but what was not changed, from the time it left my hands, were all the functions that NASA was authorized to perform and how it could perform them.

ROBBIE: Did you write the opening lines?

PAUL: I wrote them, but there were changes made in the draft. Senator Lyndon Johnson insisted that there had to be a National Aeronautics and Space Council. He wanted it to be comparable to the National Security Council, where decisions on the big policy matters could be made. The Council was established, just as Johnson wanted, but it was repealed later. House Majority Leader—later Speaker—McCormick—who had just moved enactment of the Atomic Energy Act through the Congress four years earlier—wanted a comparable Military-Civilian Liaison Committee in NASA. That was included in the Act, but was later also repealed.

ROBBIE: Why did they repeal the Council?

PAUL: The Council was organized in the Executive Office of the President, and Welsh was appointed its Executive Director. The original provision in the NASA Act called for the President to chair the National Aeronautics and Space Council, with the Executive Director as its operating head. When Senator John Kennedy became President and Senator Lyndon Johnson became Vice-President, Johnson convinced the President that the Vice-President, with his prior involvement in establishing NASA and his interest in the space program, should chair the Council. The President finally agreed.

NASA thereupon requested the Congress to enact the change in the legislation. By that time, James Webb was NASA Administrator. He was concerned that the strong, domineering Vice-President, Johnson, would, in effect, take over the space program. President Kennedy, who had appointed Webb, was similarly concerned. Webb convinced Kennedy that he, the President, should control the agenda of the Council. It would consider and act only on such topics as the President designated. This course of action greatly limited the Council's activities and, in time, it outlived its usefulness. Reorganization Plan No. 1 of 1973 finally abolished the Council and its functions abolished, effective July 1 of that year. Later, President George H.W. Bush resurrected the Council, but he did it by administrative edict and appointed Vice-President Dan Quayle as its chairman.

ROBBIE: So you went off for two weeks and wrote the Space Act. You asked Dryden "Can I do this?" He said, "Yes," and you went off and wrote it. Then what happened? What did you do with it?

PAUL: We forwarded it to the Bureau of the Budget, now known as the Office of Management and Budget, which is the executive office body that clears all legislative matters sent to the Congress in the name of the President. Before it went up to the Hill, the Bureau sent it around to the agencies for comment. The Department of Defense wanted to make sure there was a niche carved out for them, and the language finally adopted in the Declaration of Policy and Purpose of the Act did just that.

The Act states that "a civilian agency shall exercise control over aeronautical and space activities of the U.S. except the activities peculiar to, or primarily associated with, the development of weapon systems, military operations, or the defense of the United States (including the research and development necessary to make effective provision for the defense of the United States) shall be the responsibility and shall be directed by the Department of Defense."

The draft also provided that NASA could engage in international affairs. The State Department took vehement exception to that. They said the President conducts the foreign affairs of the nation. The compromise said that NASA, "under the foreign policy guidance of the President," could engage in a program of international cooperation and its peaceful application, pursuant to Presidential agreements.

ROBBIE: Did Dryden read it himself?

PAUL: Yes, indeed.

ROBBIE: Did he make any changes?

PAUL: He went over it very carefully and asked me to discuss each of the provisions with those with responsibilities in that area. He made a few changes.

ROBBIE: So the bill went to all these different agencies and everybody had their input.

PAUL: Well, not completely. The departments and agencies had 24 hours to comment. *(laughter)*

ROBBIE: Who made that rule?

PAUL: The Bureau of the Budget. They said, "We don't need to have a lot of comments." If you look at the testimony of the Congressional hearings on the bill, you'll see that Donald Quarles, who was then Deputy Secretary of Defense, complains that they didn't have enough time. In 24 hours, they couldn't get it down to the various agencies that were supposed to look at the draft bill—like Army, Navy, and Air Force. We appreciated the 24-hour limit, and we were surprised that the Bureau of Budget took that position.

ROBBIE: Was this because of the urgency of Sputnik?

PAUL: Well, that was the justification for it.

ROBBIE: Was that really the reason, or did they just sort of ram it though?

PAUL: The thinking was, "Look, we did everything we had to, we talked to everybody, let's get it moving and push it as fast as we can." The Bureau of Budget people were satisfied, we were satisfied—"Let's go!" You know—"strike while the iron is hot"—or somebody might mess it up. So, the comments on the draft bill had to be back in 24 hours. Quarles must have mentioned that five times in his testimony at the hearing.

Then the President sent it up officially to Congress. President Eisenhower sent it, and therefore it was known as the "Eisenhower Bill." The Congress had the Executive Branch's version, and now both the Senate and the House started to consider it.

ROBBIE: Were there hearings on the bill?

PAUL: Oh yes!

"The Fat Lady in The Cannon"

PAUL: Doolittle, Silverstein, and I testified at the hearings. There was a series of hearings by both the Senate and House on various aspects of the bill. During one of those, somebody asked when could we do a trajectory shot, and that was when Dr. Dryden made the big faux pas and said, "Well, it's like shooting a fat lady out of a cannon." It cost him the Administrator's position.

ROBBIE: Oh, it did?

KEN: It had that much of an effect?

PAUL: It had that much effect. First of all, they always thought he was a Republican.

ROBBIE: Who thought?

PAUL: People on the Hill, primarily—Speaker McCormick and some of the leaders on the majority (Democratic) side of the Committee. I don't know about Senator Johnson. Of course the media picked it up, and the members of Congress raised hell about his statement. They felt that he didn't have his heart in the program.

ROBBIE: What's ironic about that is Max Faget told us later that it was actually true, that putting a man in orbit was just a stunt.

PAUL: Sure! Everybody realized it—the scientists realized it—it was right. He shouldn't have said it, that's all. It was not the politic thing to say. Dr. Dryden always called things the way he saw them. He was a scientist, not a political person. And in those days, you didn't say, "Hey, they took me out of context."

At any rate, people on the Hill were very upset. They didn't think he appreciated the space program. Then, later on, McCormick learned that Dryden was a Democrat or had registered as a Democrat. He called me up to his office and informed me that he had just learned this—and more importantly "Why did someone not tell me?"—he might have been able to ease the criticism. Dryden wasn't really a political character—he had been registered as a Democrat but was not active. He'd been the Deputy Director of the Bureau of Standards before becoming Director of NACA. He started in the Bureau as a young man shortly after earning his Doctor of Science degree from Johns Hopkins.

ROBBIE: We heard that he was somewhat reluctant to move NACA into NASA just because NACA had been such a wonderful, well-functioning organization.

PAUL: He was concerned. I don't know if "reluctant" is correct, though. He was concerned that we were moving into a really big operation. We had been a research organization, rather than a development and operations organization. NACA hadn't really built large items such as weapons systems or done a lot of sophisticated contracting, and of course, that's what the Air Force was using against us in the argument—"What do they know about contracting big weapons systems the way we do?"

KEN: And there was some validity to that argument!

PAUL: There was some validity to it, but you can bring the people in who do know, and that's what happened. NASA brought in some very good people—a lot of them from the Air Force, a lot of them from the military. The then newly-appointed Director of Procurement, Ernie Brackett, was recruited from the Air Force.

ROBBIE: When you were in Washington, writing the Space Act, were you aware of a group of people coming up from Langley and having meetings in that building? A group called the Space Cadets?

PAUL: I do remember. That was one of the arguments, of course—that there were people in NACA who were engaged already in space activities and who ought to be recognized, and that Dr. Doolittle and Dr. Dryden ought to do something recognizing this work. Why wasn't it well known? NACA was an aeronautical organization. The leaders were reluctant to expose the time and effort that were being spent on space research for fear of criticism, and perhaps even a comparable decrease in appropriations.

At about this time, NACA did establish a Space Committee, chaired by a nationally known scientist, a member of the National Academy of Science who was later appointed Director of the National Science Foundation—Guy Strever. The transfer to NASA of all other agencies that were involved in space activity was provided for in the law. NASA became an agency, with NACA as its foundation, and Dr. Keith Glennan was appointed Administrator.

Working with Congress under Glennan and Webb

ROBBIE: Was Dryden bitter and upset that he didn't get to be Head of NASA?

PAUL: I think he was quite disappointed initially. He was appointed Deputy Administrator and stayed on until he died, which occurred during Jim Webb's administration. I don't think he was appreciated by Glennan as much as he was by Webb. Glennan, I think, felt that he owed him something, so he said that he wanted him as his Deputy. Dryden was an excellent advisor. He'd been Home Secretary of the Academy of Sciences and concentrated on international space matters. He worked extensively with the Russians, along with various other activities.

He was busy, but it wasn't until President Kennedy named Webb administrator that Dryden was really able to maximize his potential. Webb made it a condition of his acceptance of the appointment that Dryden would stay as Deputy Administrator. Webb was a fantastic manager and operator. Webb and Dryden—and later on, Robert Seamans—they were the triumphant troika that ran the agency.

ROBBIE: What happened with you after the Space Act passed?

PAUL: The House Space Committee wanted to write a statutory provision for a General Counsel into the act—they pushed to have that done. I said, "I don't think it ought to be a statutory provision because then it will become politicized, and I'd rather it not be politicized." NACA was a very non-political agency. The General Counsel position had not been included in the Act when NASA was organized.

Keith Glennan, who became the Director, had worked with John Johnson, who was General Counsel for the Air Force at the time. Glennan invited me to lunch and said there had been an arrangement through various sources—I gathered that it had been through the White House—leaning on him to appoint Johnson as General Counsel, so that's what he did. No one was officially named as Deputy General Counsel at the time, but I served as the Deputy with the title of Assistant General Counsel. The title was changed and I became Deputy General Counsel in 1964.

Going back to 1961, let me bring you up to date. President Kennedy had been inaugurated in January. James E. Webb was sworn in on February 14, 1961. I remember because it was Valentine's Day. I used to send him a Valentine's card every year, saying, "We're celebrating the anniversary of your being sworn in." He was the best manager I've ever seen! One of his first jobs was tackling the Congressional relations of the agency, and that was what I did when he came aboard. He appointed me Director of Congressional Relations, or Legislative Affairs. James Gleason, the Congressional Relations Director, was leaving, and Webb asked me to organize the Congressional office.

At the peak, we had it set up so that we were providing the Hill with all kinds of information and giving them a range of advice with regard to space activity—arranging for speakers in their districts, exhibits, astronauts to visit and speak, and for the Mobile Vehicle Demonstration Unit project NASA used to send around the country to the schools. They were called "Demonstration Units." Actually, they were vans equipped with working models of various NASA projects. There was usually a teacher with either a Master's or an advanced degree in science who took it to the schools and showed the kids how the various programs worked and that sort of thing. The lecturers worked the models and explained their significance. At its peak, there were more than 50 people in the Congressional office. When I took over, there were four or five.

ROBBIE: You really expanded it.

PAUL: Yes. In particular, we tried to educate the members of Congress. Various Congressmen were saying things like, "Why should I support the program? All the money's going to California." Then, when the

Mercury project began and the contract went to McDonnell in St. Louis, they said, "Everything's going to Missouri." I said, "Let's ask McDonnell how many sub-contracts they let, and what's the breakdown?" So they told me, and there were something like 3,000 subcontracting companies in 25 states involved in supplying McDonnell with what they needed.

ROBBIE: So it wasn't all going to Missouri.

PAUL: The other comment, of course, was, "Why are you putting all that money in space? You ought to use it on the ground." We tried to show them that all the money *was* being spent here on the ground. I went to our procurement people and said, "Look—let's put a provision in each contract that says each contractor has to send us a postcard showing the name and address of each company to which a subcontract was awarded, the amount of the contract, and where the contract will be performed." The first and second-level subcontractors had to do the same thing, so we'd get all these postcards and organize them by Congressional district. *(laughter)*

KEN: Aha! You would target them for the right Congressional district.

PAUL: So now when a Congressman would call us and say, "I want a speaker," or something like that, we'd say, "By the way, Congressman, do you know that $8 million is being spent in your district?" "Oh really! I didn't know that." Or when somebody made a speech on the Hill, the next day he'd get a visit with all these statistics. "Do you know, in your district alone, how many companies are involved in the space program? Look at them, and look at the people they employ. Look at the amount of money that's spent in your district. You're talking against the space program. You don't want us to tell the people out there that their Congressman is against the space program." "Oh no!"

I vividly remember a senator from my home state of New Jersey, Harrison Williams.

He raised cain about New Jersey not getting its share of the space program, and I said, "Well, we're spending $8 million to $10 million." He said, "Yeah, but that's nothing compared with other states." I said, "Look, why don't we work with your press agent? I know the newspapers in New Jersey—the *Newark Evening Star*, the *Ledger*, the *Newark Evening News*, and a couple of journals. Let's see whether we can send out a press release from your office indicating how much is being spent and see what happens."

The idea was to indicate that Senator Williams was proud of the amount of money that's spent in New Jersey on space. There are 15 companies with contracts amounting to $10 million. His reply was, "Well, if you can do anything with that, go ahead." We did, and he got good coverage. Then he wanted me to come up and appear on his television program. As you know, the Senate and House have their own television programs that they produce and send to the stations in their home districts, or their home state. I appeared with him and said afterward, "Hey, you got a lot of publicity." He said, "You're right, I won't argue," and never voted against the program. We also pointed out that there was one Congressman from Pennsylvania who voted against the program and didn't get re-elected. Now, there was no cause-and-effect, but we always pointed out that he voted against the space program and didn't get re-elected *(laughter)*. The people want the space program! So we serviced the Congress very well.

ROBBIE: I think that is a brilliant strategy! Of course the money in space is not being spent in space. When you think about it, all that's in space is a little bit of the hardware—the part that flies. The money and the infrastructure stay right here, on the ground. The way it's talked about, we're spending all this money in space, but we're not. We're spending money on Earth, for the benefit of people on Earth.

PAUL: I want to show you something. There was only one guy, Wally Schirra, who didn't help us out.

ROBBIE: Okay, this is a tiny piece of paper that says, "Mr. Dembling, apparently you are not as busy as we are." What did this mean?

Paul: The congressmen and senators would always ask us for autographed photos of the astronauts, of the original seven. Walter Schirra was the only one who always gave us trouble. He wrote that note when I sent down a whole batch of them. *(laughter)*

Robbie: In other words, the message was, "Stop wasting my time!"

Paul: I remember calling him up and I said, "Look, you're busy only because the Congressmen voted you money. If you want to continue getting money, you'd better sign those photographs and get them back up here."

Choosing the Site for the Manned Spaceflight Center

Ken: When NACA became NASA, we didn't have all the field centers that we have now.

Paul: That's right. NACA only had Langley Aeronautical Laboratory in Virginia, Ames Aeronautical Laboratory in California, and Lewis Flight Propulsion Laboratory in Ohio, plus a station at Wallop's Island, Virginia. There was also the High Speed Flight Research Center at Edward Air Force Base in the Mojave Desert, which was later renamed "Dryden" in honor of Hugh Dryden.

Webb established a three-man committee to analyze the sites proposed for a new center. The Deputy Director of Ames, Jack Parsons, headed it, and Wesley Hjornevic and Phillip Woods—a construction engineer from Goddard—completed the committee. Webb made me the Coordinator of the Site Survey. The reason, he told me, was that he didn't want to be bothered with the members of Congress or other politicians or local people calling him and saying "Take a look at our site—we meet the criteria."

Robbie: He wanted *you* to deal with that, not him.

Paul: That's right. There were 14 criteria that had been established for picking a site. The location had to be a place that was warm year-round, it had to be near a university, and it had to be near water transportation— things like that. It had to have a good urban environment, so the scientists could have a good living environment. The site survey team went out to various predetermined places. Occasionally, someone would request a visit to such and such place, and sometimes it was done because a Senator or Congressman wanted to show that he or she had some influence.

Robbie: So you sometimes made site visits not because you thought the site was so viable, but just to make a Senator look good.

Paul: That's right. But Webb didn't want to make that decision. He could just as easily say, "Dembling is making all those decisions, so call Dembling," and that's what happened. I recall a couple of occasions that we did it. Senator Russell wanted us to go to Georgia to check on an island off the coast. Then someone asked us to come up to Massachusetts, insisting that it was just as warm up there as it was down in Georgia! Because the President was from Massachusetts, we went there.

Another one I remember, because it helped us later on, was a Louisiana Congressman who said, "I know we don't meet the criteria that you set up. However, I'm running against a man who is criticizing me and saying I've been in Congress for 25 years and can't even get the team to come down into my state." He asked us to visit two cities—Victoria and Bogalusa. He said, "I've been supporting NASA," and the Administration thought that was a good justification, so the team went there and he was very appreciative. It was primarily for what is now known as a "photo opportunity." The team appeared with the Congressman on television and explained the purpose of their visit, saying they'd come into the state and the area at the invitation of the Congressman. He was pleased, and later that kindness redounded to NASA's benefit.

Let me explain. In the NASA Act, we had a provision for what were called "excepted positions."

That meant they were excepted from the civil service pay schedule, or positions exempt from the Civil Service pay structure, above and beyond what could be paid on the regular scale. The NASA Act authorized the agency to appoint 125 scientific, administrative, and engineering personnel without regard to Civil Service laws, and to fix their compensation not in excess of the highest rate of Grade 18 of the Classification Act.

So we requested125 and that's what we received in the organic NASA Act. This was a real breakthrough. The Civil Service Commission had strenuously objected to this authority.

But we were locked into that 125. Later on, when we needed an increase, I went up to Congress to see what I could do. At the time, Congress was considering a personnel bill. When I went to the Senate, Senator Monroney, who was chairing the committee that had jurisdiction, told me they had completed action and had just sent it over to the House. I then went over to the House side. The chairman of the committee was Congressman Murray from Tennessee. He wasn't available, and his deputy, the ranking member of the committee, had taken charge. That was our friend, Congressman Morrison from Louisiana.

ROBBIE: What luck—you knew he was already favorably disposed!

PAUL: Well, I hoped so. I told him what our problem was, and he said, "You know, I have that bill right here, sitting on my desk. I've got to take it into Committee conference." He's sitting down at the desk and he says, "Come over." *(laughter)*. He took out his pencil, and he said, "What number were you thinking about?" So I said "425," and he repeated it—425? I said "425"—so he wrote in "425."

ROBBIE: He just made the one into a four.

PAUL: That's exactly right! He said, "I'll carry it into conference tomorrow, and we'll carry it through." Then he turned around, remembering that I was the one who sent the site team in to his district, looked up to me, and said, "I never forget my friends!"

When I got back to headquarters, Webb said, "How'd you do?" I said, "Well, Senator Monroney told me it was over on the House side, so I went over to see the Acting Chair of the Civil Service Committee." Webb said, "How'd you do there?" I said, "He changed it." "What did he change it to?" I said, "425." He said, "425!! What in the hell are we going to do with 300 additional positions? Whatever prompted you to ask for that many?" I said, "I didn't think he was going to give me the whole 425!" He said, "What did he say?" I said, "He didn't say anything. He just changed it. He turned around to me and said, 'I never forget my friends!'" *(laughter)* "Okay," Webb said, "Now the slate's clean. He picked up your chit. The slate's clean—you start anew from now on. He's given you everything." By the way, NASA had no problem in using those 425 positions.

Robbie: That's just what Eilene said happened with McCormick. When she suggested changing Agency to Administration, he immediately called his secretary and said "Everywhere the word Agency appears, change it to Administration."

Paul: Right. We were the only agency that had *anywhere near* that number of excepted positions.

Back To Being a Full-Time Lawyer: General Counsel for NASA

Robbie: How did your role evolve once NASA established Johnson Space Center?

Paul: By late 1963, Congressional relations were fully operational. We had people covering each of our main committees—namely the Senate and House Space Committees and Appropriations Committees, and the sub-committees of our main Committees. Staff people and members of Congress knew who to call if they needed something from NASA, so I went back to being a full-time lawyer again. Jim Webb had promised, when he asked me to take on the congressional job, that as soon as I had it well organized and on a plateau, he'd return me to the office of General Counsel. I was Deputy General Counsel and worked a lot with Webb on all kinds of things.

ROBBIE: What issues were you up against then as Deputy General Counsel?

PAUL: I worked in the procurement area and the contracts area, and sat on some of the Source Selection Boards, which determined rankings of proposals submitted by contractors. I also chaired the NASA Board of Contract Appeals, which decided disputes between contractors and NASA. I also served as Vice-Chairman of the statutory NASA Inventions and Contributions Board, which oversaw the title to space-related inventions. In addition, I served on the U.S. delegation to the Legal Committee of the United Nations, where I participated in drafting the various space treaties. Finally, I worked on the agency's management administrative problems. Webb had asked me to review all appointments to NASA for conflicts of interest, for example.

Later, I was named General Counsel and continued to work quite closely with Webb. He wanted me to sit in on meetings with him. There didn't have to be a legal problem involved—he often asked me to join him when he had a visitor he didn't want to meet alone. I also reviewed, commented on, and edited memoranda that he prepared. These were often replies to questions from President Johnson. Webb had me look at them, even when they involved topics unrelated to space or law. I recall one that he sent to the President on some proposed taxation. The President, recalling Webb's service as Budget Director, wanted his views, and Webb asked me to review and edit it.

When it was decided there ought to be an integration contractor to integrate the various stages of the boosters on Apollo, he wanted to pick somebody that could do that job. He wanted an organization set up near NASA Headquarters, preferably across the street or something like that, for ready and continuous access. Most people considered that the Aerospace Corporation was the only company that could really do the job, so he called in the Head of the Corporation. This guy—I think his name was Ingstrom—came to NASA Headquarters and Webb wanted me to sit in on the meeting with him. Webb outlined exactly what he wanted done. Ingstrom said, "No, we don't operate that way. Here's the way we operate." Webb didn't like that and they argued back and forth, and after a while, Webb thanked him and said that he'd be in touch. When Ingstrom left the room, Webb said, "He's not going to dictate to us how to do our project. *We're* going to dictate how were going to do our project." As a result of that conversation, Aerospace did not get the job. Webb called Fred Kapel, who was then heading AT&T. He asked Kapel to visit with him.

Kapel came down to Washington, and again Webb said to me, "How about sitting in?" So I did. He told Kapel, "I don't want you to run a nonprofit organization; I want you to run a profit-making organization. We recognize that we're a partnership. It's going to be a fair organization, but I want the best engineers and scientists you have in the Bell Laboratories." Kapel didn't want to do it initially, claiming that they didn't do that kind of work. He said, "I can't give you my good people out of Bell Laboratories. We've got a terrific ongoing organization." Webb was very persistent, and eventually Kapel gave in. They had several conversations, and finally Webb convinced him that he could put together a company that would perform very well. The organization that came about as a result was BellComm. They not only did the integration, but also went on to do a lot of other things and had as many as a thousand people working for them at one time. It became a very big company.

Ken: And that was what he was comfortable with, in terms of his management style?

Paul: That's right. He figured he gave the Aerospace Corporation a chance and they'd lost their opportunity, and probably future ones as well, by trying to tell Webb how it was going to be done. It was a "take it or leave it" approach that turned Webb off.

Webb was quite an organizer and an excellent manager. He also understood the Congress and how to operate on the Hill. Let me give you an example: on January 27, 1967, the calamitous fire that cost the lives of astronauts Grissom, Chafee and White happened at Kennedy Space Center. Coincidentally, on the same day, the nations of the world signed the Outer Space Treaty in London, Moscow, and Washington. President Johnson hosted a reception for the representatives of the countries who had signed the treaty at the State Department earlier that day, and it was during this reception that Webb was notified of the disaster at the Cape.

He quickly asked those of us who were at the reception to join him back at his office. During that meeting, he asked me to start preparing for an investigation, and over the next two days—

Saturday and Sunday—we formulated preliminary plans. He was determined that NASA would conduct its own investigation, and contacted the President and convinced him that the White House should not establish any investigative body. He and I then met with the Chairmen of House and Senate Space Committees and sub-committees to forestall their calling any hearings. Webb promised that NASA would conduct a complete investigation and would report to the Hill periodically on its progress. He promised that, at its conclusion, he'd come to the various committees and explain how the investigation was conducted, what witnesses had been called and what had been learned, and that he'd disclose what actions had been proposed and recommended so that such a tragedy would never happen again. On those conditions, the Hill agreed.

Webb was very much on top of everything. He did things that led people to think he was impulsive, but he sorted it all out very clearly. He might call half a dozen people around the country and get their ideas about something, then walk into a meeting and sit and talk about it, but he didn't tell people, "I've checked with six or seven people and we've discussed this." I don't know why he never disclosed those conversations. He'd just say, "Why don't we do it this way?" and people would say, "Jeez, there he goes, shooting from the hip again."

ROBBIE: So he didn't explain himself.

PAUL: He explained himself about everything else, but he wouldn't tell people that he checked around before making a decision. Let me give you another example where he brought me in. This involved Mr. Atwood, then head of North American Aviation, later part of the Rockwell Group. The North American-Rockwell Company was the contractor for the Apollo capsule, in which the fire occurred in January of 1967. Webb asked Atwell to come to Washington to discuss what he planned to do, and asked me to join him at all the meetings. Webb's position was clear. He told Atwood that while the company had performed in accordance with the contract, had delivered on schedule and was due a bonus, he could not face Congress or the American people and say that NASA was giving a bonus to the company in whose capsule three astronauts lost their lives. The company could sue or take any action it wanted, but he was determined. He also took the opportunity to pressure Atwood to reorganize—something Atwood eventually did. The meetings were tense, hostile, confrontational and long and, since I was included, Atwood asked his General Counsel to join also, so we sat mostly as silent observers, commenting only when asked to do so while the two principals went at it for hours.

I was pleased that I enjoyed his confidence. Some years after he had retired, during one of my luncheons at his home, he brought up the subject of the many papers and letters that he still had in his files. He had previously donated his papers to either the Truman Library, or the Johnson Library, but he had reserved the sensitive ones—reports, confidential letters from people, that sort of thing. He asked me to go through them and see if they were still sensitive. If not, he wanted to donate them.

Mostly, they were papers from people doing things for him on the side. Say, for instance, an official from Lockheed, or some other company, was going into the Ames Laboratory. Webb would ask him to take a look at such-and-such a project while he was there, and let him know what he thought. The person would report back with candid opinions. Webb had monitors of that kind—people who were really knowledgeable in the field, but were visiting the center for some other reason. He'd say "While you're in the center, by the way, why don't you take a look at such-and-such a project? You don't have to tell them that you're looking at it for me—just tell them that you're interested in taking a look at it. Let me know what you really think about it."

ROBBIE: So did you let those papers go on to the library?

PAUL: Some of them, where the person wasn't alive anymore, or it wasn't sensitive.

Professor Henry Lambright, from Syracuse University, wrote a biography of James Webb—a very well-done book. About two years ago, I wrote a review of it for the *NASA Alumni League Newsletter*. The title of the book is *Powering Apollo: James E. Webb of NASA*, published by Johns Hopkins University Press. I recount a couple of stories about Webb too, personal stories, for example, of how he arrived at the budget figure for the Apollo program and how we brought it in under that amount. Everybody always said, "If NASA can do it, why can't the military and everybody else?" It became a benchmark for estimating.

What happened was this: Brainard Holmes was head of the Apollo program at the time we were scheduled to testify before the Appropriations Committee. We knew that the Chairman, Albert Thomas, always asked, "What are the runout costs on this program?" In other words, what were the costs for the whole program?—because agencies usually justify appropriations for one year. As you may know, Congress normally appropriates funds for one year.

So, Webb asked Holmes to come up with an estimate for the total cost of the Apollo program. In effect, he said, "Don't tell me how much you need this year, tell me how much the whole Apollo program is going to cost." Holmes came up with a figure of $10 billion and Webb said, "You know, I think that's too low—you better come up with another figure," so he came back with $13 billion. "Okay, $13 billion is the one we're going up to the Hill with." During testimony, the Chairman of the House Appropriations Committee asked Webb what was the run-out cost and Webb said, "$20 billion."

Robbie: Why did he say $20 billion?

Paul: Going back in the car, Brainard Holmes asked the same question: "Where did you get $20 billion? We gave you $13 billion!" Webb said, "I put an 'Administrator's discount' on it." What Holmes didn't know was that Webb had read a book by Peck and Sherrod—a book called *Weapons Program Acquisitions* that analyzed the acquisitions that had taken place in the military weapons program. It covered the original estimates and the amount of the overruns. Webb had really studied that book, since NASA was beginning to engage in large, costly programs. Nobody else knew that. I knew because we'd discussed the matter. He said, "You know, every large program overruns by such-and-such," and he used that figure to add on to the $13 billion estimate, and that's how he came up with the $20 billion figure. So, when he went up to the Hill, he said, "$20 billion." Brainard and the other guys thought he'd acted impulsively and he just said, "I put an administrative discount on it!"

Robbie: Did he get the $20 billion?

Paul: Yes, over the years. People like Senator Proxmire were always criticizing all the other agencies for not knowing how to estimate. Proxmire would taunt the military and other agencies, "How come NASA said $20 billion for the Apollo program for 10 years and they brought it in for $20 billion? How is it that can they estimate, and you can't?"

Ken: And the actual results *were* close to that.

ROBBIE: What did Webb do in his retirement?

PAUL: He remained very active. He was interested in what was going on in the space program. We got him involved in an oral history project that we called the Glennan-Webb-Seamans program. We felt there were a number of people who were going to be leaving us and the stories would be lost if they weren't captured. We were anxious to have the people involved in the NASA program be interviewed to record their thoughts and comments. We didn't look to iron out differences. We decided to leave that for the scholars. We just felt that we ought to organize and get something done, so we raised some money. The Air and Space Museum had a small history group that undertook the project. Willis Shapley headed the Glennan-Webb-Seamans program at first, and when he went back to work at NASA, I took it over. Martin Collins of the museum acted as Executive Director of the program. He was our liaison—the person with whom we worked on a regular basis.

After NASA

ROBBIE: When did you leave NASA?

PAUL: I left in November of 1969. We'd put the men on the Moon in July, and I figured I'd done my job. I was planning to go into the private sector. Elmer Staats had been appointed Comptroller General of the United States and needed a General Counsel, because his previous General Counsel, Bob Keller, had been appointed Deputy Comptroller General. I guess Webb had told him that I was leaving NASA and he should try to recruit me. Staats had been Webb's deputy when he was Director of the Bureau of the Budget under President Truman.

Staats had asked me to visit him after work, ostensibly to obtain my views on NASA's relations with the GAO, which the Comptroller General heads. I didn't know he was looking me over for a position. One visit led to several, and he offered me the position of General Counsel. I had a private offer at the time, but I decided to take the job at the GAO.

I thought it would be fun, educational, and challenging—an opportunity to see the entire government's operation from the legislative side. I was there from November 1969 to November 1978.

ROBBIE: And what are you doing now?

PAUL: Oral histories *(laughs),* and working on selective projects. I originally organized the Washington office for a firm—Schnader, Harrison, Segal and Lewis. The firm is a regional one, based in Philadelphia, with more than 200 lawyers.

ROBBIE: Are you actively practicing law?

PAUL: I was practicing full time until I retired officially a couple of years ago. They asked me to stay on as Senior Counsel, so I did. I consult with the younger people, do a project here and there.

ROBBIE: So you've had, really, four interesting careers. You were in the War Department, working for NACA, and then NASA, and then as a GAO, and now in the law firm. It's amazing what you can do in one life!

PAUL: All my jobs have been interesting, educational, challenging, and self-fulfilling in different ways.

ROBBIE: When you look back over those four careers, where do you feel like you've made your greatest contribution?

PAUL: It's hard to say. There were many events at NASA. Drafting the NASA Act was certainly a highlight in my life. Working as a member of the United Nations Legal Committee and participating in drafting the various outer space treaties is something I rank as a significant contribution, and heading NASA's Congressional Relations for close to three years during its formative stage is also something I'm proud of. Similarly, there were events at GAO that were significant.

NASA Then and NASA Now

PAUL: Before we close, I did want to mention was that I served from 1964 to 1969 as a member of the U.S. delegation to the United Nations Legal Subcommittee.

ROBBIE: Along with Eilene? Was this the same committee she was on?

PAUL: Same one, and it was in that connection that the Outer Space Committee drafted the Outer Space Treaty. That was a *good* treaty. I wrote a very long article detailing the process of arriving at the agreement. I kept notes during the entire negotiation, so I had available to me each of the proposals of the various delegations during the years we were trying to agree. I thought there ought to be a legislative history of the Outer Space Treaty, so I wrote a very complete article on the subject. Then, at the 30th anniversary of the Treaty last year, I was asked by the organizers of the International Institute of Space Law Colloquium in Italy to summarize the significant negotiating points during the formulation of the Outer Space Treaty. I did so in a paper presented at the Colloquium in Turin, Italy, during the International Aeronatique Federation.

ROBBIE: You know that we also interviewed Guy Thibodaux, Max Faget, and Paul Purser. It's so fascinating to hear the stories from different points of view! Listening to them, you would think that all of NASA was located in the Pilotless Aircraft Research Division at Langley—that the whole space program was driven from PARD.

PAUL: Well, I do think that probably was the birth of it—if you had to say, "Where was the seed?"—PARD was really where it was.

ROBBIE: And to hear them talk, Headquarters was just something that they had to deal with sometimes, but when we interviewed folks who had worked in Headquarters, their perspective was that everything was driven from there. One of the things Tibby, Faget, and Purser talked about was how they "bootlegged" almost everything they did. They would just decide "We're going to do something" and start doing it, and the money and official sanction would show up eventually.

PAUL: Well, the approach that I liked the best was the Abe Silverstein approach. He wanted a new propulsion laboratory at Lewis and he finally had Headquarters approve it. Congress then appropriated money for the laboratory, but they didn't fund the entire lab at one time. What he wanted was two segments—he wanted a place where the scientists would work, but he also wanted administrative space. He wanted the administrative space and the scientific laboratories in the same building.

Partial funding was provided, and the assumption was that he was going to build the scientific portion first, but he didn't. In fact, he built the administrative part first. [*laughter*] He's a scientist and you'd think he'd build the scientific part first, but no. Why? "Because," he said, "Hell, now they've got to give me the money for the scientists. I wouldn't have gotten the administrative offices the other way!"

He and Webb had differences, too. Abe did something once and Webb told him he shouldn't do it. Webb thought it ought to be done some other way, and finally Webb said, "This is it. I'm not taking any more excuses, this is *it*!" He sent him a telegram saying that, and Silverstein shot back a telegram saying something to the effect, "I bow to superior authority, rather than superior rationale." At Abe's retirement, Webb read that telegram, saying, "You see what I had to put up with?"

Abe, who had been serving as Deputy Director of the Lewis Laboratory in Cleveland, was originally named the Head of the Apollo program. In fact, he was the one who named the program "Apollo." When he was asked to run the program, he started bringing up all these fellows from Lewis—Harry Finger, George Low, and Ed Cortwright—a whole series of people who were known as "Abe's boys"—all very able people. And Abe went to Webb and said, "I want this kind of authority in order to run the Apollo program," and Webb said, "No, you can't have it." Later on, when George Mueller was heading the manned space program, he assumed the authority that Webb had refused Silverstein. Webb later said, "I'm sorry I turned Abe Silverstein down—you know these guys are taking that authority without asking for it anyway." *[laughter]* When he didn't get what he thought was necessary, Abe just said, "OK, I'm out of here. Let me go back and be the Director of the Lewis Flight Propulsion Laboratory," which he did.

ROBBIE: Thibodaux told us about how he wanted to build a rocket laboratory so that he could actually build some spherical rockets he was designing. Rockets were supposed to only be built at Lewis, but he wanted to do it at Langley, and so he went to Gilruth and he said "I want to build this but there is no official charter for it." And Gilruth said, "Well, I can write checks for $999.99 without anyone else's authorization, so if no one component is going to cost more than $999.99, then we can do it." So he just wrote this long series of checks and it got done. *[laughter]* Basically, they bootlegged it.

PAUL: I remember at NACA when all this documentation wasn't necessary. After a discussion with a Center representative, you'd say, "Okay, we have an agreement." That was it, that was final. I didn't realize how strong that was until after I left, and Neil Hosenball, who'd been my deputy, took over as General Counsel some years later, and he said to me, "I ran into a very interesting situation at Ames. Art Freeman called up and said that someone in Headquarters wanted to change some policy that they'd been following for years. He called me to have it "checked with Dembling." Neil said, "Well, Dembling's left already." He said, "Oh, because that's the policy we had agreed to with Dembling." And that was years before. Art Freeman was the Administrative Officer at Ames. If you had an agreement, even an oral one, it was binding. That was true at NACA Langley, Ames, and Lewis. An oral deal, over the phone—that was it!

ROBBIE: That's what Thibodaux said, that when it quit being fun was when it quit being run that way.

Paul: Well, it became larger, that's right. At NACA, we had 100, 150, people at Headquarters office. Everybody knew each other, and everybody knew what they were doing.

Robbie: But even after JSC got big, those guys could still call up Purser, who was running the place, and say, I need "X" and he'd say, "I'll take care of it," and when it became more complicated than that, it wasn't fun anymore.

Paul: That's right. Everything was done on an oral basis—you could call somebody, and the response wasn't, "I've got to confirm this in writing," or some other excuse. He said, "Okay, I'll take care of it," and it was taken care of. It was an interesting organization. NACA was so different from anything else that I've ever read about or experienced. You had the culture of research, but the agency was also working very closely with industry. It was an interesting combination, and a lot of that culture carried over to NASA.

People ask, "What's the difference between a military organization and NASA, and why was NASA so successful?" The big difference is that NASA had the capability, in-house, to do everything that they contracted out, and the military organizations didn't. In fact, we used to have to fight with the people in the Centers. They'd have a contract and when the supply came in, they'd pull it all apart to see whether it was built the way they wanted it built, or if maybe they could make some improvements on it.

Robbie: Because they could have done it themselves in the first place.

Paul: They could have done it themselves in the first place. Exactly! They perhaps didn't have the volume capability, but they certainly had the capability of doing whatever they were asking their contractors to do, and they recognized good ones. The Germans down in Huntsville would say, for example, "We want RCA to build it. We know they'll come back with a very good product," and we'd say, "You can't do that, you got to open it to competition." "OK, compete, but make sure that RCA gets it!" *[laughter]*

KEN: How would you advise NASA today?

PAUL: I serve on the Board of Directors and I'm General Counsel of the NASA Alumni League. One of the things we're wrestling with is just that question—what should the role of NASA be?

Some of the questions include what kinds of large programs should NASA engage in? Should they be manned space programs? This is always a question. Can you do what you want to do, or what you need to do, without involving humans? How much should robotics come into play? What's the future? Should we be developing space stations? The International Space Station is moving along. That was what faced NASA at the end of the Apollo program. I think NASA moved into the space shuttle program because it was the best one that was around, not necessarily the best one for NASA at the time. It seemed a way of moving forward and continuing the large programming concern. Some feel that there shouldn't be one big program that requires a lot of money. It commits NASA too much in one direction, from which a change is not easily possible.

KEN: Well, Paul, this has been a very interesting and illuminating interview.

PAUL: It's fun remembering those exciting times and events! It was an extraordinary time in our nation's history and it was a privilege to have been part of it.

ROBBIE: You are a goldmine of information—you played a formative role in the development of NASA, and you've had an incredibly varied career. And you articulate it all so well, with lots of fascinating detail. It's been a wonderful interview. Thank you!

EPILOGUE

TO THE STARS

SOME 70 YEARS LATER, we may find it difficult to imagine that the seeds of today's burgeoning global space program were planted so unobtrusively by the "Space Cadets," "Nakker Nuts," "Young Turks," and others who were unleashed to do their thing in the 1950s and 1960s. For many of us latter-day "Space Cadets," space exploration has proceeded much more slowly than we might have liked. However, *Space Stories* offers a new perspective on how much has really been accomplished. Only yesterday, it would seem, NACA had begun to lay the groundwork for NASA, and for a long time, the American and Soviet space programs were the only game in town. Today, many states and nations are rushing to build spaceports, and the new "space race" is among entrepreneurs who made their original fortunes in cyberspace, and now seek to extend their domains outward.

Today, the highway to the stars lies before us. As we take it to unknown destinations, let us always remember those who pointed the way.

Made in the USA
Middletown, DE
05 November 2026

36545527R10173